John Milton Scudder

On the Use of Medicated Inhalations

In the Treatment of Diseases of the Respiratory Organs. Third Edition

John Milton Scudder

On the Use of Medicated Inhalations
In the Treatment of Diseases of the Respiratory Organs. Third Edition

ISBN/EAN: 9783337183912

Printed in Europe, USA, Canada, Australia, Japan

Cover: Foto ©berggeist007 / pixelio.de

More available books at **www.hansebooks.com**

ON

THE USE

OF

MEDICATED INHALATIONS,

IN THE TREATMENT OF

DISEASES OF THE RESPIRATORY ORGANS.

BY

JOHN M. SCUDDER, M. D.

AUTHOR OF "A TREATISE ON THE DISEASES OF WOMEN," ETC.

WITH

AN INTRODUCTION TO THE SUBJECT,

BY

W. ABBOTTS SMITH, M. D., M. R. C. P.

PHYSICIAN TO THE METROPOLITAN FREE HOSPITAL, LONDON.

THIRD EDITION.

CINCINNATI:

WILSTACH & BALDWIN, PRINTERS,

1874

PREFACE.

The subject of direct medication in diseases of the air passages is beginning to attract the attention of physicians, as a field likely to be worked with advantage. It would doubtless have received more attention, but for the fact that it was made a specialty by ignorant persons, and extensively advertised as effecting cures, where the common means were of no avail.

It is evident that much good might result from the judicious use of local applications to diseased surfaces, if these could be used with anything like the degree of definiteness that attends their employment elsewhere. And it is the object of this monograph, to show that this can be done with the improved apparatus now employed.

As the study of such a subject necessarily requires years, and the united effort of a number of observers, in order to develop it, I need not offer any apology for such imperfections as may be discovered in the present work. It is written to draw more attention to the subject, and to present the practitioner with the results of three years study, as a basis for the commencement of the practice.

PREFACE.

The author wishes to be distinctly understood, as claiming
for inhalations only the position of adjunct means in many
affections, and a principal means in but one or two. Insisting
that it must be combined with a judicious general treatment
if we desire to obtain the best results. But with such treatment, it will many times relieve certain symptoms, and cure
certain structural diseases, that stand in the way of the patient's recovery.

As DR. MADDOCK well remarks: "No one deprecates
quackery in every shape and form more than ourselves, but
we must say, that when the generality of practitioners are
confessedly incapable of combating the ravages of pulmonary
diseases, it is neither consistent with reason nor humanity, to
expect that the public will stand supinely by, and see the in-
roads of disease unchecked or unalleviated, and leave their
friends or relatives a prey to misery and despair, without
making one effort to arrest its progress or cheer the mind by
giving trial to new remedies or modes of treatment, especially
if supported by reputable testimony. Narrow-headed men,
with narrower minds, may whisper obstacles and superciliously
condemn a new proposal, and pronounce it useless, without in-
quiry or any opportunity of seeing or judging of what they
advise others to reject: but the abuse or discountenance of such
members of the profession can avail but little, and only proves
that such persons must be contracted not only in their ideas,
but regardless of the advantages of their patients or the
advancement of their profession. The realities, however, of
inhalation are now by far too numerous and too well attested
to be put down by contumely, or even by indifference; and
we doubt not but that from the number of authentic instances,
which we almost daily receive, of its applicability to the

treatment of pectoral diseases, that such an amount of evidence will soon have been amassed, as to shake the belief of
the most inveterately sceptical, for in the investigation of
truth the illustration of its principles are never insulated;
and however one manifestation of its presence may be kept
from receiving general acceptance by the efforts of prejudice,
others will present themselves with a constancy which shall
render their distinctive character as incontrovertible as their
real existence."

As an introduction to the subject, the little work of W.
Abotts Smith, M. D., of London, is republished. No better
argument could be employed to recommend inhalation as a
means of cure, and to stimulate research in this direction.

98 West Sixth Street, Cincinnati.

I.

ON THE INHALATION OF GASES AND MEDICATED VAPORS.

BY W. ABBOTTS SMITH, M. D., M. R. C. P.

So much has recently been said and written in extra-professional circles on the subject of inhalation as a remedial agent, and so evident is the fact, that many persons look upon it as a novel mode of treatment, that it will not be altogether uninteresting to commence this general description of inhalation by showing the opinions of some of the most ancient Professors of Medicine concerning it, in order to prove that it possesses high claims to antiquity.

In their descriptions of the treatment of catarrh, coryza, and cough, Galen, Actius, Rhases, Ægineta, and other ancient authorities recommend the inhalation of the fumes of various substances, which are ordered to be ignited, so that the person affected may draw the vapor arising from them into the throat and lungs through a funnel. Haly Abbas and other celebrated Arabian physicians, who, at a later period, were the sole possessors of medical learning, also speak favorably, amongst other remedies, of the inhalation of the vapors of vinegar, camphor, and other substances   In asthma and phthisis, again, similar fumigations were recommended; and we find one old writer, Avicenna, speaking in high terms of the inhalation of the vapour of pine fruit, a plan which has recently come into vogue in some parts of Germany.   References may also be met with in some of the older treatises on Medicine to fumigations with certain mineral substances, reduced to the condition of vapor, in the treatment of secondary syphilis.

It would be easy to multiply these instances of the value which was set upon inhalation, not only by the older medical writers, but also by those who have lived in later times.   Yet, with all this, there are many who would have others believe that inhalation is a new method of treatment, and who, through combined ignorance, avarice, and love of notoriety, hold out specious hopes of recovery to persons whose cases are unfortunately beyond human aid.   So much, indeed, for the claim of novelty which has been put forward in respect to inhalation.

Still, there is much in the practice of inhalation when rightly used, to make it worthy of employment in many cases of pectoral disease ; and I   believe that most medical men will agree with me that it is the indiscriminate use of inhalation, and not a judicious resort to it, when it is suitably indicated, that is to be deprecated.*

---

* I have heard the opinion advanced, but as I venture to think, very illogically, that because inhalation has been unfortunately adopted by empirics, this method of treatment should be discountenanced by the Medical Profession.   I must confess I cannot see the force of this argument.   If there be anything in inhalation, it is our duty, in the interests of our patients, to give them the benefit of it, and to show to what extent

Like the Turkish bath, and many other agents of undoubted remedial value, inhalation occupies a lower position in the therapeutic scale than it would do were it not for the discredit which has been brought upon it by the extravagant views which have been advanced by its supporters, and which have led to consequent disappointment.

Inhalation was confined to fumigations with the vapors of various vegetable, and occasionally mineral, substances, until towards the close of the last century, when the discoveries which were made of oxygen, and other gases, induced several distinguished physicians and chemists, including Priestly, Beddoes, Cavendish, and Humphry Davy, to anticipate most favorable results from the inhalation of different gases in the treatment of consumption and other diseases. For the purpose of practically carrying out the views of the pro- moters of the system of Pneumatic Medicine, as it was called, an institution was founded at Bristol, by Dr. Beddoes, with Humphry Davy (then just beginning his illustrious career) as superintendent. Capacious reser- voirs were constructed for the reception of large quantities of oxygen, carbonic acid, carburetted hydro- gen, and other gases, and patients flocked thither in considerable numbers from all parts of the country to avail themselves of Dr. Beddoes' treatment. Various circumstances, amongst which may be comprised the too sanguine ideas of the founders of this establishment, and the costly nature of the apparatus employed, led to

it may be rendered useful in the alleviation or cure of disease ; but to re- linquish any remedy because some ignorant person has laid hands upon it, and from his want of knowledge, or some other cause, has pretended tc discovered merits in it which it does not posess, seems to be opposed to com mon sense the ultimate abandonment of the project, but not before researches had been conducted on a sufficiently extensive scale to show that much good might be derived from the inhalation of some of the gases employed, especially of oxygen, if some plan could be devised by which they could be generated in a less costly and more portable manner, so that patients could use them at their own homes, instead of their being compelled to undertake long and hurtful journeys. Those of our readers who desire additional details of the history of Dr. Beddoes' " Medical Pneumatic Institution," will find much interesting information in " Beddoes' Letter to Erasmus Darwin, M.D., on a New Method of treating Pulmonary consumption," published in 1793, in Davy's "Chemical and Philosophical Researches,"* written by that author whilst he was superintendent of the Medical Pneumatic Institution, and published in 1800, and in the valuable Memoir of Sir Humphry Davy, edited by his brother Dr. John Davy.

The impediments just referred to, thrown in the way of administering oxygen by inhalation, appear to have baffled later writers on pulmonary affections. Laennec,

---

* During his residence at Clifton, ne ir Bristol, where the Medical Pneumatic Institution was situated, Davy experimented upon most of the gases then known. Dr. Druitt, in his very useful " Surgeon's Vade Mecum " (seventh edition, p. 702), points out the important fact that Davy, in conducting some experimental researches on the properties of nitrous oxide, found that its inhalation mitigated the pain of cutting a wisdom-tooth, and from other circumstances connected with its inhalation, he threw out the hint that as it appeared to be " capable of destroying physical pain, so it might probably be used with advantage during surg.cal operations.' From this discovery of Davy's and that of cotemporaneous observers that the inhalation of ether, pure, or medicated with conium or some other vegetable sedative, allayed the irritation in spasmodic asthma and whooping-cough, may be traced the inestimable boon of Chloroform, as an anæsthetic.

in his "Treatise on Mediate Auscultation, and Diseases of the Heart and Lungs," remarks that "no means could seem better calculated to combat the dyspnœa which arises from increased want of respiration, in spasmodic asthma, than the inhalation of pure oxygen," but, as he adds, the difficulty of promptly procuring it is a complete bar to its employment.

This difficulty no longer exists, for oxygen can now be generated in sufficient quantity, and at such a moderate cost, as to allow of its general use; and which is of still greater importance, the trials which have been made of it by various observers show that it possesses remedial powers of no small value.

The largest and most complete series of experiments with oxygen gas both upon animals and human beings, are those which, during several years past, have been conducted by my esteemed friend M. Demarquay, of Paris, in conjunction with M. Leconte. These careful and trustworthy observers have recently embodied the results of their researches in some reports presented to the Academy of Sciences of Paris; and their conclusions concerning the physiological and therapeutical properties of oxygen so closely coincide with those of other experimenters, that they may be advantageously epitomised in the following description of the effects of oxygen when inhaled.

The first series of experiments, which comprised a large number of observations, had relation to the action of oxygen on animals only.* In these experiments,

* In reference to experiments conducted upon animals, physiologists have of late been attacked, in no measured terms of reproach and abuse, for the alleged cruelties committed by them in the prosecution of their inquiries. Without entering upon a lengthy defense of scientific men against such charges, it may be observed, that a physiologist, working for the im-

it was ascertained that dogs are able to respire, during a long period, as much as thirty or forty litres of oxygen, or even more (a litre, it may be remarked, being equal to 1·76 English pints), without injury, the only apparent result of the inhalation being to render the animal more lively and to improve his appetite. In some instances large wounds were made in the axillary region, and the animals were made to respire oxygen when these wounds were healing. It was then observed that they became brightly injected with arterial blood, and that a transparent, serous fluid was effused at their surface; and, further, that as the inhalation of the gas was continued, numerous petechiæ, or ecchymoses, were produced. In order to ascertain whether similar results would be produced by the injection of oxygen into the venous system, a series of injections into the external jugular vein were made, and the same effects were produced as when inhalation was practiced. The principal results shown in the summary of the researches of MM. Demarquay and Leconte into the effects of inhalation of oxygen upon animals are:— 1. That after death, caused by the constant respiration of this gas, the muscular system of the animals subjected to it was found to be in a very turgescent condition. 2. That, contrary to the opinion advanced by Broughton, an

provement of science, and the ultimate benefit of the human race, inflicts less pain and torture in the whole course of a year than many a sportsman would cause in a few days' shooting, indulged in merely for the gratification of his own pleasure. A rabbit (one of the animals most commonly selected for physiological experiments) would suffer far less pain at the hands of a physiologist, who, whenever practicable, would most likely render it insensibe by the administration of chloroform, previously to the commencement of his experiments, and who would put it to a speedy death afterwards, than the same animal would suffer at the hands of a sportsman, if, with broken leg or mangled body, it had crept away into a thicket to meet a tedious termination to its misery.

early inquirer into the physiological action of oxygen, the venous and arterial symptoms retained their natural color. 3. That, contrary to the assertion of Beddoes, no organ, however vascular it might be, was ever found to be the seat of either inflammation or gangrene; and 4. That the muscular system assumed a peculiar rosy hue. MM. Demarquay and Leconte next proceed to describe the action of oxygen upon man. In the first place, when it is applied locally to the surface of wounds, by means of special caoutchouc apparatus it gives rise to a slight sense of heat and tingling, without pain. In the course of a few hours the suppuration becomes diminished in quantity and consistency, and the granulations assume a greyish color, and appear to be smaller in size. After the oxygen has been removed, they again become red and turgescent; and, if the application of the gas be renewed for several hours daily, for some succesive days, more or less inflammatory action is induced. One of the most remarkable effects of the oxygen, topically applied, is the rapid manner in which it modifies the conjestive redness commonly present at the circumference of a wound; and, in this manner, the redness surrounding ulcers of the extremities, and the injection of the skin remaining after eczema, may be readily removed. This property of oxygen, when applied locally, will, doubtless, at some future period, be turned to beneficial results in the treatment of indolent and other forms of ulcers and wounds, as also to that of certain affections of the skin.*

* While briefly refering to the original uses of oxygen, the fact may be here stated, that the tunica vaginalis has been injected with this gas without inconvenience, hydrocele having in one instance subsequently undergone cure

It is, however, with the effects of oxygen, when inhaled, that I purpose to deal in the present paper. I must, therefore, pass on to this point.

Demarquay and Leconte, with their pupils and friends, found that they could readily inspire a dose of from twenty to thirty litres without any inconvenience resulting from the inhalation; and no accident has occurred from its use by a large number of patients during an extended period. The daily inhalation of twenty to forty litres, continued for the space of a month or six weeks, gave rise to a moderate feeling of warmth in the fauces and chest, occasionally accompanied by a slight headache. As a rule, the pulse at first increases in the number of beats, but in some persons the pulsations become less frequent; and the inhalation is generally followed by increased appetite and strength—the improvement of the former being frequently very remarkable, while the assimilatory powers become more vigorous. These changes are not, however, so well marked in patients who have been previously worn out by chronic illness. A singular alteration is observed in wounds, whether recent or old, after patients have inhaled the oxygen for several successive days; these become red and turgid, and suppurate much more freely than they had done before the inhalation was commenced. This peculiar action of oxygen explains why its inhalation is attended with such unsatisfactory results in the last stage of tubercular phthisis. Patients in this condition derive considerable benefit at the commencement of the practice of inhalation, but the inflammatory symptoms soon become more intense (as in the case of external suppurating surfaces), and these are followed by abundant expectoration and more urgent cough, and death would

soon occur if the inhalation were persevered in. In
this way, doubtless, disastrous consequences have resulted where persons, ignorant of the physiological
properties of oxygen, have directed its indiscriminate
inhalation, and discredit has been thereby thrown upon
oxygen as a remedial agent; but this circumstance,
instead of tending to depreciate the real value of this
remedy, simply furnishes a proof of the necessity which
exists for proper professional advice before the adop-
tion of inhalation, as, indeed, before following any
other method of medical treatment.

The therapeutical applications of oxygen are numerous, and it may be safely resorted to, excepting
when certain contra-indications, which will be enu-
merated further on, are present. Oxygen is particularly serviceable in the treatment of cases of disease
attended by an anæmic or chlorotic condition, in cases
of debility, and in affections which exercise a depressing effect upon the system, such as, for instance,
diphtheritis, diabetes, and the secondary and tertiary
forms of syphilis; in affections of this nature, if the
age and general state of the patient are favorable,
the inhalation of oxygen is soon followed by an improvement in strength and spirits, and often greatly
increased appetite. The lips and surface of the body
assume a more healthy color, greater vitality is
manifest, and much of the nervous irritability previously present disappears. During the course of inhalation we must, however, especially inquire into the
condition of the internal organs, and, in fact, of the
whole body, because, as has already been stated, suppurating surfaces, under the stimulating effects of the
oxygen, become so greatly modified in their character,
that inflammatory action is eventually set up, unless

the case be carefully watched. At the same time, the
mere existence of a sore or a wound is not, of itself,
sufficient to contra indicate the employment of oxygen :
indeed, on the other hand, this remedy may be not unfrequently used with great advantage in the treatment
of certain kinds of sores or wounds, which are charac-
terized by the absence of sufficient vitality, and which
consequently remain stationary, or heal only very
slowly. The action of oxygen is much sooner mani-
fested in the young than in old persons.

In the foregoing description of the physiological and
therapuetical effects of oxygen, I have purposely followed the reports presented to the Academy of Sciences
of Paris by MM. Demarquay and Leconte, because their
researches have been more extensive than those of any
other experimenters. Other observers, however, both
in this country and on the continent, have arrived
at similar conclusions, and the interesting discussion
which followed a paper by Dr. Richardson, read at a
recent meeting of the medical society of London, as
reported in the February number of the "Medical
Mirror," shows that the subject of inhalation is one
which at the present time attracts a large amount of
professional attention.

The efficiency of oxygen is undoubted in most cases
of disease attended by debility, and diminution of the
red corpuscles of the blood. In asthma it sometimes
acts like a charm in removing the difficulty of breathing, and restoring the patient to a healthy state. In
the later stages of consumption it is inadmissible, unless
resorted to only occasionally and for a short period, for
the reason which has been already given, but when
inhaled earlier in the course of the affection, it may be
productive of considerable benefit.

The inhalation of oxygen is contra-indicated in affections accompanied by much fever; in deep-seated inflammatory disorders and visceral diseases; in most affections of the heart and large vessels;* in neuralgia occurring in plethoric individuals; and when the hæmorrhagic diathesis is present.

The difficulties which formerly existed in connexion with the practice of inhalation, such as those of obtaining gases pure, and also at any time or in any place, are now almost obviated. With respect to oxygen, a portable and very useful apparatus in which it can be generated, and from which it can be inhaled, was lately exhibited before the Medical Society of London. It can be procured of Messrs. Garden and Robbins, of Oxford Street, by whom is also sold an oxygennesis-powder, with which the gas can be easily and promptly prepared.

Besides oxygen, numerous gases have been employed therapeutically, and some of them possess valuable properties.

The vapor of iodine, one of the best disinfectants known, may be occasionally inhaled with advantage by persons suffering from scrofulous affections, enlargement of the tonsils, sore throat, ozæna, and similar disorders. The readiest mode of inhaling it is from a box with a perforated lid; or a small quantity of the tincture of iodine may be added to some hot water, so that the patient can inhale the iodized steam. Dr. Murray, in his "Treatise on the Influence of Heat and Moisture," and some other authors state that

---

* In Cyanosis, the inhalation of oxygen, properly employed, is productive of great benefit. The patient improves in general health, the color of the face and lips becomes more natural, and many of the signs of imperfect oxygenation disappear under this mode of treatment.

they have employed iodine in the gaseous form, with
much benefit to the patients. Dr. Murray says, that
he has always observed improvement of a temporary
kind, at least, in the condition of the patient; the
cough becomes diminished in violence and the expectoration easier, while the patient sleeps better. Other
observers have not arrived at such favorable conclusions as this writer, but there appears little doubt
that the inhalation of the vapor of iodine forms an
efficient method of administration in cases where
iodine or its preparations are indicated, such as hyper-
trophied tonsils, etc. One of the latest observers on
this point, M. Simon, states, in the "Union Medicale,"
1861, that out of twenty-eight phthisical patients, seventeen derived positive benefit from iodine inhalations,
both as to general and local symptoms, and four might
be considered as cured. Iodine has been employed in
this form also in cancer, but without any decidedly good
results.

Another gas which has been highly spoken of by
some writers in the treatment of phthisis, is chlorine
gas, which has ever been credited by some with the
cure of this affection. Louis, in his "Researches on
Phthisis," states that he studied the action of chlorine
on upwards of fifty consumptive patients, without any
sufficiently successful results to warrant him in speaking so favorably of it as some other authors had done.
It is probable that the discrepancy on this point, as
upon most of those medical questions about which
there is a diversity of opinion, is principally due to
the fact that the cases experimented on by various
observers differed either in degree or even in kind.
Louis himself appears to admit that the inhalation of
chlorine gas is efficacious in chronic pulmonary catarrh,

(a point on which numerous authorities are agreed), although not in tubercular phthisis. It may be employed whenever direct stimulation of the mucous membrane of the air-passages is indicated.

The other gases which have been most frequently used for inhalation are nitrous oxide, carburetted hydrogen, and carbonic acid. The first named has been greatly extolled in the treatment of asthma. Carburetted hydrogen was employed by Davy, who thought that it was possible that by various combinations of nitrous oxide, "we should be in possession of a regular series of exciting and depressing powers, applicable to every deviation of the constitution from health" (vide "Davy's Chemical and Philosophical Researches"). Carbonic acid, administered by inhalation, when diluted with atmospheric air, has been recommended by various writers. According to Vogler, (Deutsche Klinik, 1859), the respiration of the carbonic acid, diluted with air, given off from the springs at Ems, is attended with more or less irritation, so that it is likely to be beneficial only in cases where stimulating treatment is indicated.

Air, either moist or dry, medicated with various remedial agents, has been employed therapeutically from the earliest periods of medical science, and the mere enumeration of the different substances which have been thus used would constitute a formidable looking list. The action of these medicated vapors is, of course, almost analogous to the effects produced by the same remedies when administered by the mouth, and they are consequently indicated in a similar manner, according to the individual requirements of the case, and to the special indications for stimulants, anti-spasmodics, or sedatives. I must, therefore, content myself with a brief mention of the principal remedies which are, or have been, administered by inhalation.

Arsenical fumigations were held in high esteem by the ancients in the treatment of asthma, bronchitis, catarrh, and some other affections of the air passages and lungs. The preparation of this mineral which they employed was a more inert substance than the arsenic employed in modern times. It was almost identical in composition with yellow orpiment, and is supposed to have consisted of sixty-two parts of arsenic and thirty-eight of sulphur, according to Klaproth (*vide* "The Sydenham Society's Translation of the Works of Paulus Ægineta," vol. i., p. 480). Arsenic is seldom employed in inhalation at the present day, but cigarrettes which have been steeped in a solution of arsenious acid are occasionally recommended to be smoked by asthmatic persons. Dr. Leared has published some cases in which signal benefit was obtained from smoking these cigarettes.

The ethereal preparations, especially sulphuric ether, are sometimes useful as sedatives and antispasmodics, when inhaled. A few whiffs of air, containing a small proportion of chloroform, will be found very serviceable in allaying or removing the irritable cough of spasmodic asthma.

The inhalation of the vapor of hot water containing camphor, conium, belladonna, hyoscyamus, lobelia, stramonium, and other vegetable sedatives, is attended with very beneficial results in all diseases of the chest, in which there is much local irritability and troublesome cough.

Creosote and tar have been recommended for use

in a vaporized form, when stimulating remedies are indicated.

Nitrate of potash fumes have long been employed in a similar class of cases; and those of hydrochlorate of ammonia have been more recently well spoken of in cases where a stimulating plan of treatment is necessary. In using nitrate of potash, pieces of blotting-paper, previously soaked in a saturated solution of the salt, and then dried, are burnt upon an earthenware plate; the fumes are soon diffused throughout the room, and their beneficial effects are, in many cases rendered evident, as Dr. Watson observes, when writing of the value of this remedy in asthma, "in clearing out the passages, and gradually opening the air-tubes." The inhalation of the fumes of hydrochlorate of ammonia is recommended in pulmonary catarrh by Paasch; they may be easily generated by pouring a little hydrochloric acid into a watch-glass placed in a saucer containing some liquor ammoniæ.

The different balsams and gum-resins which possess expectorant properties may be volatized by heat, and their vapors inhaled, with good results, in asthma and bronchitis. Those most commonly employed in this manner are, the balsam of tolu, balsam of Peru, benzoin, and storax.* Their efficacy may be increased by the addition of a little spirit.

* As a direct proof of the antiquity of the practice of inhalation, the following passage may be quoted from the writings of Paulus Ægineta who lived at an early period of the Christian era :—" To be inhaled for a continued cough : storax, pepper, mastich, Macedonian parsley, of each one ounce ; sandarach (an arsenial preparation), six scruples ; two bay-berries ; mix with honey ; and fumigate by throwing them upon coals so that the person affected with the cough may inhale the vapor through a funnel. It answers also with those affected by cold in anywise." The chief objection to this prescription is, that there are too many ingredients contained in it.

The benefit obtained by the inhalation of moist medicated vapors is attributable, in some measure, to the soothing effects of the steam of the hot water in which the various remedies are dissolved. I am in the habit of recommending patients suffering from congestive asthma, bronchitis, and analogous pulmonary affections, to inhale at intervals the simple vapor of heated water, at different degrees of temperature, according to the nature of the case, without the addition of any medicinal agent. The warm, moist vapor is most agreeable to the patient, and seldom fails to afford immediate relief. The expectoration becomes freer, the cough is less frequent, and the dryness and irritation of the air-passages are entirely removed. The inhalation of hot steam is very efficacious in the treatment of hay-asthma or hay-fever, as I have already pointed out, in a communication to one of the medical periodicals, on the subject of this troublesome affection.

There are numerous methods by which the patient may be enabled to inhale the steam, as for instance, from a basin or jug containing hot water, or by breathing through a sponge dipped in hot water, and partially wrung out. Several forms of apparatus have also been devised for this purpose. The most efficient apparatus for the inhalation either of simple steam or of medicated vapors is that which is known by the name of Nelson's Inhaler; it is constructed of earthenware, and, in addition to its complete adaptation to the purpose for which it is intended, possesses the triple recommendation of cleanliness, portability, and cheapness.*

Until a very recent period, only such remedial sub-

---

* This useful apparatus is manufactured by Messrs. Maw & Son, of Aldersgate Street.

stances could be used in inhalation as could be administered in the gaseous form, or volatilized into vapor, so as
to allow of their being drawn into the lungs during the
act of inspiration. During the last few years, however,
the discovery of the laryngoscope has led to the closer
investigation of the diseases of the larynx and air-passages generally, and various new methods of applying
local medication have been devised. Of these the most
ingenious, and, at the same time, the most important,
are those which have for their object the minute subdivision of different remedial substances, so that they
can be inhaled, when in solution in the form of spray.
The first apparatus invented for this purpose was that
of Dr. Sales-Girons, but it has been since modified and
improved upon by several individuals, particularly
Lewin, of Berlin, and Siegle, of Strasbourg. The
instrument invented by Sales-Girons in 1858, is so constructed that the medicated fluid is forced, by the
agency of compressed air, through a tube having a
very small opening against a metal plate. At this
point the stream of fluid is checked, and it becomes
divided into fine spray, (to which the term of pulverized, or atomized, has been applied), and in this condition can be inhaled by the patient. A modification of
this has been introduced by Bergsen, of Berlin, whose
apparatus consists of two glass tubes, having capillary
openings at one end—these two ends being placed
almost at a right angle with each other. The more
open end of the perpendicular tube is immersed in the
medicated fluid, and, as the compressed air is forced
through the horizontal tube, the air in the perpendicular one becomes exhausted, and the medicated solution
then rises in it, and, when it arrives at the capillary
opening, is dispersed in very fine spray by the force of

the compressed air passing along the other tube. The principle upon which this instrument acts is familiarly illustrated by the perfume-odorators which have been introduced into general use during the last few months, and which are, in fact, merely an adaptation of Bergson's ingenious invention to common purposes. Siegle's atomizer is constructed on a similar principle, with the substitution of steam for compressed air as the means of dispersing the medicated liquid. It has been doubted whether solutions, thus divided into minute spray, reach the lungs, but the fact that they do so has been incontestably proved by numerous observers. In one striking case, experimented upon by M. Demarquay, it was shown that the inhaled fluid reached the trachea. The case was that of a woman who had a tracheal fistula, and in whom, after she had inhaled the spray from the pulverizer, the introduction of chemical tests through the tracheal opening proved the presence of the various substances which had been inhaled, thus clearly proving that, at any rate, the spray which had been drawn during the act of inspiration had reached as far as the fistulous opening. Numerous other proofs of the entrance of the medicated spray into the air passages, and even the lungs themselves, have been brought forward. The most satisfactory proofs of this fact are the practical results which have been obtained by the administration, in this manner, of different remedial agents, and the marked benefit which has resulted from their employment in hæmoptysis, asthma, bronchitis, whooping-cough, and some other pulmonary affections. The principal medicines which have been used in this way, are nitrate of silver, alum, tannic acid, iodide and bromide of potassium, bicarbonate of potash, tincture of the sesquichloride of iron, sulphate of zinc, acetate and hydrochlorate of morphia, tincture of opium, the extracts of hyoscyamus and conium.*

With respect to the inhalation of dry air, as distinguished from that of moist vapors, two plans of treatment demand notice, viz., that by compressed air, and that of the inhaling tube, invented by Dr. Ramadge many years since.

Of the former I have no personal knowledge. The idea of employing this novel mode of treatment is stated in the "Gazette Medicale de Paris," to have originated in the circumstance that several of the workmen who were engaged in sinking the foundations of a bridge, and who happened to be affected with various chest diseases, were comparatively free from the symptoms of their complaints when they were at work in the caissons sunk below the surface of the water. Establishments at which patients are subjected to the effects of condensed air have been formed at Paris, Lyons, Montpelier, and some other continental cities; and the promoters of this plan of treatment describe it as more or less successful in cases of catarrh, chronic bronchitis, asthma, anæmia, etc. The pressure used averages from $1\frac{1}{5}$ to $1\frac{2}{5}$ atmospheres. I must own that I am very skeptical as to the benefit alleged to be derived from treatment by compressed air, particularly as numerous cases have been put on record both in this country and abroad, of the production of paralysis, and other serious consequences, in laborers who have

* Since the above given paragraph was written, the author has seen a report (Lancet and Medical Times, Feb. 25th) of the meeting of the Medico-Chirurgical Society, at which a paper by Dr. Mackenzie "On the Inhalation of Atomized Liquids" was read. Dr Mackenzie and Dr. Gibb, both of whom are able observers, were agreed as to the value of inhalation in the treatment of certain affections of the air passages and lungs.

remained subject for too long a period to the effects of compressed air. Those who are desirous of complete information concerning this plan of treatment will find a long paper upon it, by Dr. Sandahl, in Schmidt's "Jahrbuch" for 1863.

The inhaling-tube, after having crossed the Atlantic, has been re-introduced into this country, and made widely known through the medium of monster announcements in the daily journals, by an individual whose sole and very slender claim to any knowledge of the treatment of pulmonary affections is based on the mere circumstance of his having brought out an English reprint of a book written by an American author. This tube is constructed in such a manner that air can be easily drawn through it into the lungs, but is not so quickly expired. The normal balance between inspiration and expiration is thus re-established, and in suitable cases for this kind of inhalation, the chest becomes gradually expanded, so as to resume more natural dimensions. Dr. Ramadge says that is a valuable adjunct to other treatment in consumption and asthma.

II.

THE APPARATUS FOR INHALATION.

A vague idea of the benefit of inhalation in the treatment of diseases of the air passages is present with nearly all physicians; but it rarely assumes any tangible form, because of the dearth of information in the text books. Occasionally we have a spasmodic manifestation of it in the use of the vapor of water or decoction of hops, tansey, etc., in acute disease, inhaling the fumes produced by burning nitrate of potash in asthma, or of resin, Canada balsam, balsam of Peru, tar, etc., in bronchitis and phthisis. But though occasional benefit results, the attempts are so crude they are soon given up.

The oldest method of employing inhalations, was, to vaporize the material employed on a hot iron. Thus any fluid might be employed, or any solid, like the resinous or balsamic agents, brought into a condition that they might be inspired. Another method, in the case of fluids, was, to place them in a vessel and by the aid of a hot iron, brick, or stone, to produce the necessary amount of vapor. Though the methods were crude, I have seen very marked benefit result from their use. The employment of the vapor of water would rarely fail to soothe irritation in disease of the larynx and bronchii, and when this induced contraction of the muscular tissue and consequent difficulty in respiration, would give temporary relief until it could be rendered permanent by the use of other means.

Volatile agents were used in many ways. Many years back, I recollect seeing them put in a common wine bottle partly filled with hot water, and the vapor inhaled as it passed from the mouth of the bottle. Others would employ a coarse sponge. Pressing it out of hot water, the medicine was sprinkled upon it, and breathing through it the patient obtained the vapor of the medicine, and a moist air which was not unfrequently of full as much benefit.

Many forms of apparatus are employed at present, each being thought preferable by its inventor. We need describe but four of these, as they will fulfill every indication. As the remedies employed may be divided into two varieties, *volatile* and *non-volatile*, two forms of apparatus will be indispensable.

The volatile agents are vaporized by heat, and a majority of them at a temperature less than that of boil-

FIG. 1.

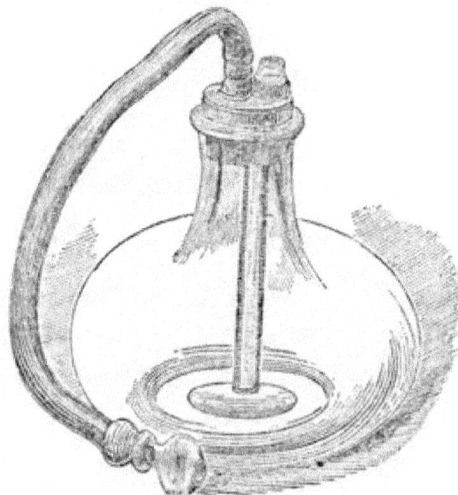

ing water. It requires then a container to receive the
fluid, and a tube to conduct the vapor to the patient.
Fig. 1, represents such an instrument. The flask is
made of flint glass, and will usually bear an elevated
temperature. The cork is perforated by an opening for
the attachment of the rubber tube, and for a smaller
glass or metal tube for the admission of atmospheric
air, as the patient inhales the vapor. Nelson's Inhaler,
spoken of on page 22 is also a good instrument.

In place of this, I have used the apparatus represented
in Fig. 2. It consists of a tin cup, perforated at the

FIG. 2.

bottom and a three quarter inch tube inserted and sol-
dered. To this, is attached two or three feet of rubber
tubing, which is terminated by a mouth and nose-
piece as represented in the wood cut. The cup con-
tains two cross wires to hold the sponge, which should
be coarse and open. In using this apparatus, the
sponge is pressed out of hot water and put in the cup,
or it may be wet in the cup, and the medicine then
sprinkled on it in sufficient quantity. The cup being
placed upon the floor, a chair, or, if more heat is
required, upon the stove, the patient inhales the vapors
as they arise. In country practice, it is well to have
the cup made the size of a tea-kettle, or tin teapot lid,

so they can be turned over these vessels if it is desirable to inhale the vapor of water. This apparatus recommends itself, in that it is simple, cleanly, and cheap, being easily manufactured wherever a tinner can be found, at a cost of 75 cents.

The volatile agents, however, form but a small portion of those we should like to use, and we must, therefore, have some apparatus for conveying to the air passages the remedies themselves. Such an apparatus we now have in great perfection, evenly distributing any fluid, no matter what its character. They are known by the generic name of *Atomizers*, inasmuch as they minutely divide the fluid, that it may be inhaled. Though there are a number of these, it will be necessary to describe but three.

The simplest instrument is that known as " Elsberg's Nebulizer," which consists of two hard rubber tubes pointed at the extremities, the openings being small, and so hinged that they can be placed at right angles, the openings being immediately opposite, as in Fig. 3.

FIG. 3.

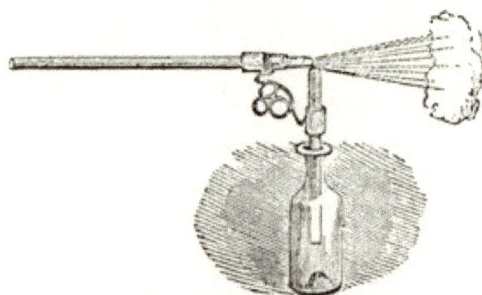

One arm of the apparatus, being placed in the medicated fluid, blowing through the other, causes the fluid to rise in the tube, and it is carried off in a fine spray. Rimmel's *Rafraichisseur*, which is the same in principle, has been employed for some years for distributing perfumes, and may be purchased quite cheap. The principal objection to this method is, that it requires a second party, and the breath cannot but prove offensive to many patients.

The second form of apparatus, consists of a cylinder in which works an air tight piston, like the barrel of a syringe. Fluid being placed in it, is forced through minute openings in the nozzle, as a delicate spray.

Fig. 4.

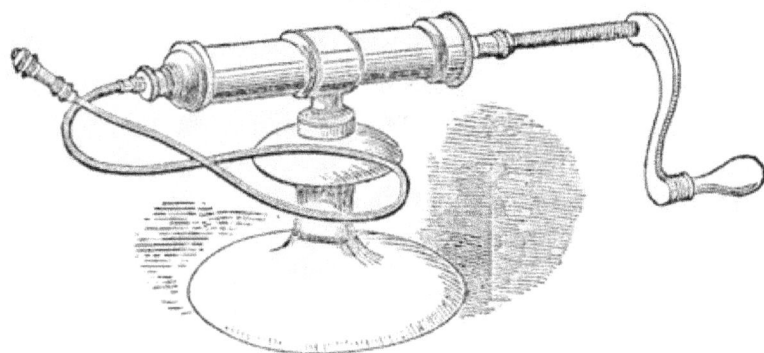

Fig. 4, represents the instrument of M. Sales Giron, which I have used in my practice with excellent results. When inhalations are much used, I have no doubt they will be manufactured by our hard rubber manufacturers, at a price to bring them within the reach of all.

The apparatus of Dr. Mackenzie is a very good one. The piston is drawn back by a wheel and rack at its upper part, and is forced down by a circular spring which surrounds the cylinder. The apparatus is filled with liquid by a funnel in its top, and all the spray, except that which is inhaled, passes back into the apparatus. He claims the following advantages for it.

" 1. Its simplicity, requiring only a few turns of a handle to set it in operation. 2. The extremely fine state of subdivision which it effects. 3. The uniform pressure exerted. 4. The fact that the waste liquid returns into the apparatus.   5. The case with which it can be taken to pieces and cleaned."

Two or three other varieties are used, but the principle is the same. In the French instrument, the piston is forced down with a screw, and the liquid forced out with very great pressure. Four or five turns with the wheel is sufficient to afford a jet of spray for as many minutes.

With these instruments, the medicated liquid is minutely broken up, and the spray inhaled with the air. In fact, the division is so fine, that it will pass wherever the air goes. The temperature of the fluid may, to some extent, be adapted to the case; but the general impression, whether the fluid is used hot or cold, is that of coolness.

FIG. 5.

The third form of apparatus is that of Dr. Seigle, and is far preferable to the others, for its simplicity and because it is automatic. The best reason for preferring it, however, is, that its price is such as to bring it within the means of any patient, as it is furnished through the druggists for $5,00, and its construction is so simple, that it is readily operated by any one.

The instrument consists of a little kettle, into which

is screwed a fine cork perforated with a horizontal tube, in which there is a fine opening. Placed at right angles to the horizontal spout, is a vertical tube, which dips down into a small cup containing the medicated fluid. As the steam issues from the horizontal tube, it causes a vacuum in the vertical tube, and the medicated liquid rising up becomes mechanically incorporated in the steam, and is blown off in the form of a minutely divided spray. The dilution of the medicated liquid which takes place is very slight, as the conversion of a drachm of water into steam will take up three drachms of medicated liquid. The temperature of the steam is lowered by the incorporation of the liquid, so that at the end of the cylinder, it has only a temperature of seventy degrees. The instrument represented in Fig. 5, is manufactured by Mr. Max Wocher of this city.

"In this apparatus there is, of course, no current of cold air. The amount of liquid taken up varies, that· is, it depends on the amount of heat applied, on the height of the column of liquid, etc. This is not an important defect; but when it is desired to take up a definite quantity of liquid, the author uses the following apparatus: A graduated glass tube, about eight inches high, has from its lower part a fine piece of tubing, which is bent round and up again, and then extends about an inch horizontally, and ends with a minute opening. In the vertical portion of the fine tubing, there is a stop-cock. The small apperture of the tube is bent at right angles to the tube from the kettle, and as the liquid emerges, it becomes incorporated in the steam. By means of the stop-cock, the amount of liquid which passes from the tube can be regulated, so that the same amount can always be taken up at the same time."

I have employed this instrument in every case where inhalations could be used with the least prospect of advantage, and feel justified in recommending it to the profession. There can be no doubt, but that the medicated fluid is carried through all parts of the respiratory tubes. This diffusion is an objection in the treatment of some affections, as we would like to restrict the remedy to the diseased surface. Still as we learn more of the action of medicine upon the body, we will care less about its concentration, and more about selecting the appropriate remedy.

I also employ a modification of Seigle's apparatus in city practice. It consists of a small boiler, with two openings at the top; one for the insertion of the glass tube in a cork, the other closed by a cork, for filling the boiler with water. This boiler is placed on a Musgrave's, or other small gas heater, which furnishes the heat. It answers an excellent purpose, and is cheaper.

Fig. 6.

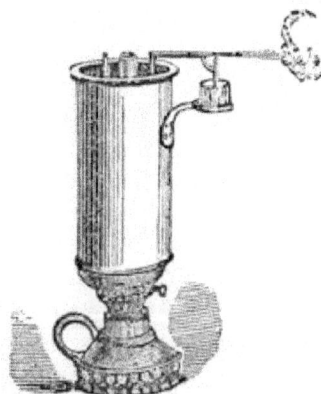

Another apparatus of like character is constructed upon the plan of the Kerosene Lamp Heating Company's boiler. Coal oil is employed, and in these days of high taxes and dear alcohol, the use of coal oil in these instruments is an object. I think it decidedly the best atomizer in use, not only for its cheapness, but for its safety, and the fact that we can get a much stronger jet of steam from it than from the original Seigle apparatus: this is sometimes very desirable. Fig. 6 represents the apparatus, which may be obtained of all prominent druggists, or by addressing J. G. Henshall and Co., Cincinnati. The price is $5,00, securely packed for shipment.

Dr. Mackenzie, who has experimented to a considerable extent with atomized liquids, thus sums up the results, in a paper to the Medical Times and Gazette:

" Leaving Demarquay's rabbits out of the question—it having been shown by Claude Bernard that as those animals in their normal state breathe through the mouth, the conditions are not physiological; and by Fournie that any solution (not atomized) injected into a rabbit's mouth passes into the lungs—there are: 1. Demarquay's and Brian's experiments on dogs. 2. His (Dr. Mackenzie's) on pigs and dogs. 3. An experiment performed by Demarquay, in the presence of numerous witnesses, on a woman with a tracheal fistula, in which it was shown that the inhaled liquid penetrated to the trachea, though there was a great obstruction at the upper opening of the larynx. This experiment, which had been previously unsuccessfully performed by Fournie, has since been repeated by Lieber, Schnitzler, and others, with results similar to those obtained by Demarquay. 4. The fact first shown by Bataille, and since by Moura-Bourouillon, the author, and others, that after the inhalation of a colored atomized solution, the sputa remained tinged long after the employment of the laryngoscope could detect any traces of the material used.

"On the one hand, there were an immense number of positive proofs of the penetration of atomized liquids, on the other hand there were a few experiments performed with negative results. It was scarcely necessary to remark that any experiment might be performed—the most simple chemical test employed—in a manner to insure failure. But a few experiments of this sort could have little weight against the mass of evidence on the other side.

"The author stated that the greatest benefit from this system of therapeutics might be expected, and had resulted in bronchitis, asthma and hæmoptysis. He brought forward twenty-two cases treated between October, 1863, and January 1864. There were ten cases of bronchitis, six of phthisis, two of hæmoptysis, three of asthma, and one of hooping-cough. The author did not believe that in phthisis the treatment would have a positively curative effect, but was beneficial in cutting short intercurrent bronchitis. Of the twenty-two cases detailed, only two were unable to make use of this curative process. Of the ten cases of bronchitis, eight were cured, one relieved, and one obtained no benefit.

"The average duration of the time required for curing these cases, though most of them were severe and of long standing, was only fifteen days and a quarter. The shortest time was six days (a severe case, No. 4); the longest forty days. The duration of treatment was not in proportion to the severity of the disease, one mild case requiring twenty-eight days to get well.

"Of the six patients laboring under consumption, two were unable to use the inhalations on account of the irritation which they caused. Of the remaining four cases, whilst the physical signs did not undergo any material alteration, the local symptoms (expectoration, pain, and cough) were greatly relieved. The general health was much improved in two cases. Nos. 11 and 15; slightly in a third, and not at all in a fourth. In two cases of hæmoptysis, one severe, the other slight, the atomized liquids rapidly stopped the bleeding. In three cases of asthma—one very severe case, which had obstinately resisted the ordinary treatment—this system of therapeutics soon gave relief. In one case of hooping cough (in an adult) the inhalations gave immediate relief. In one case of hooping cough (in an adult) the inhalations gave immediate relief, and quickly effected a cure. The author stated that during the past year he had used atomized liquids in more than eighty cases of diseases of the lungs, and that he had found the plan of treatment no less successful than was detailed in this paper. The various instruments referred to in the communication were brought before the society, and likewise diagrams illustrating their action and method of employment."

III.

ON THE THERAPEUTICS OF INHALATIONS, WITH FORMULA.

It seems almost like a work of supererogation to describe the topical action of remedies upon mucous membranes, and yet the subject of topical medication is so new, that many physicians have not thought upon the subject sufficiently to apply the general principles of therapeutics with which they are familiar.

The mucous membrane of the air passages does not differ materially in its structure, or its vascular and nervous supply, nor yet in its function, from other parts of the general mucous lining of the body. Hence, what we know by experiment and experience with regard to the topical action of remedies upon mucous structures, is applicable here. And as the principal difference between the skin and mucous membrane is in its epithelium, it will be found that remedies act in a very similar manner upon both. Thus, each one may start with a very fair knowledge of the action of remedies, and he will rarely find himself mistaken.

Of course, it is of much importance to determine the exact condition of the part affected. As regards its circulation, dryness or increased secretion, tension or relaxation, etc., giving the same care in the examination as we would to diseased surfaces that we can see and wish to medicate. Auscultation gives us this information very accurately, and in a short time the careful observer will have no difficulty in determining these points. The general symptoms are also an important guide, as the mucous membrane does not differ materially in its general condition from other parts of the system. Thus we would never expect to find a tense, dry mucous membrane, with a relaxed and flabby skin, nor a dry, irritable condition of the air passages with a soft and feeble pulse.

Whilst the general and local disease resembles each other to such an extent that the one may serve to some extent to define the other, they should be disassociated in the treatment. The general treatment will have reference to digestion, the state of the blood, excretion and innervation. It generally resolves itself into the use of the most feasible means for improving nutrition of textures, and the more active the nutritive processes become, the better are the chances for complete recovery. Thus a phthisis is many times successfully treated by the use of bitter tonics, iron, and a nutritious diet, care being taken to keep the secretions free. Added to this, the restoratives, preparations of phosphorus, cod-liver oil, etc., with out-door exercise and pleasant society, and the general treatment is as near perfect as we can expect to render it with the means at our command.

The local treatment is directed more especially to the relief of irritation, change of secretion, and to restore tone and strength to the mucous membrane. The remedies used, might be appropriately classified as *relaxant or emollient, narcotic, stimulant, tonic,* and *astrin-*

gent, though we will find it somewhat difficult to properly arrange all of them under these heads.

Relaxant or *emollient* inhalations are employed when there is irritation with dryness, or a tenacious mucus that is raised with difficulty; a condition similar to that for which we would apply a poultice externally. The vapour of water is the type of this variety, and will be found useful in the treatment of all acute inflammations in the first stage. A decoction of hops, of poppy heads, of German chamomile, or tansy, is sometimes better than water alone. If it is desirable to produce great relaxation, I usually employ an infusion of the herb lobelia, or a small portion of sulphuric ether. If there is a tenacious secretion difficult of removal, the addition of common vinegar proves valuable: in these cases, I usually order vinegar and water equal parts.

These inhalations may be employed by any mode that the fluid can be vaporized, and even the crude methods first described give excellent results. Of course the apparatus of Seigle will be better than these.

By a *narcotic* inhalation I mean one that relieves irritation of the nerves distributed to the mucous membrane, and thus checks cough. The name is not a good one, as these remedies rarely produce that deadening of sensibility that might properly be called narcosis, yet as it is the only term by which the action above named is known, I use it for want of a better. The term *sedative* has been employed, and might still be used were it not now so generally applied to those remedies that control the circulation.

By a narcotic inhalation we understand, then, a remedy which, when locally applied, relieves irritation of the part and checks cough. They are sometimes used for this purpose alone, but are more frequently added to other inhalations to effect a double purpose. Thus, added to water, or simple infusions, they are emollient-narcotic; to tonic preparations, tonic-narcotic, etc.; and as the relief of irritation and cough is one of the prominent objects of treatment they are extensively employed in this way. It must not be supposed, however, that they are the only remedies for the relief of a cough, for many times we obtain more speedy and better results in this respect from the use of stimulants, tonics and astringents.

This class of remedies divide themselves into the volatile and non-volatile, the first being employed with the apparatus described in Figs. 1 and 2, the second class either by vaporization or by the apparatus for atomizing fluids. Almost every narcotic or sedative may be employed in this way, and upon exactly the same principles that would govern their local application elsewhere. The following will be found good formulæ for the volatile agents:

No 1.—℞ Hydrocyanic acid dilute, f℥ij; wine of ipecac, paregoric, aa f℥ss; tincture of conium, f℥ij; rose water, f℥xij. Half an ounce of this may be inhaled three or four times a day.

No. 3.—℞ Cyanuret of potash, gr. viij; tincture of ipecac, tincture of lobelia, aa f℥vj; tincture of stramonium, f℥j; rose water, ℥iv; M. Inhale a teaspoonful every four hours.

No. 4.—℞ Acidi hydrocyanici diluti, min. xx; tincturæ hyoscyami, tincturæ lupuli, aa f℥j; aquæ calidæ, ad. f℥viij: M. (Tanner.) Employed in phthisis, ulceration of the larynx, etc.

No. 5.—℞ Acidi hydrocyanici diluti, min. xv; spiritus chloroformi, f℥iij, aqua bullientis, f℥viij: Mix. In laryngitis, œdema of the glottis, etc. (Tanner.)

4

No. 6.—℞ Tincturæ chamomilla, f℥ss ; morphia sul. gr. x ; æther, alcohol, aa f℥j : M. Use two drachms for an inhalation.

No. 7.—℞ Tinct. conii, tinct. strammonii, aa ℥ss; spiritus chloroformi ℥j : Use two drachms for an inhalation.

These formula might be increased, but sufficient has been given to illustrate the mode of combination in most frequent use. When employing fluids with an atomizer we need not be so particular as to their form. Tinctures added to water or proof spirit give the most eligible preparations, unless we except the aqueous extracts rubbed down with water. We do not combine opium with the other narcotics, as it is antagonistic in its action to most of them. Indeed, it will be found best to use each agent separately. The following will be found good preparations.

No. 8.—℞. McMunn's elixir of opium, f℥ij., decoction of hops, ℥iv. Mix. Of this half an ounce may be employed as an inhalation. It acts as a stimulant to the respiratory organs, and hence is never employed when there is dryness, but answers an excellent purpose, when there is relaxation of the bronchial mucous membrane with increased secretion.

No. 9.—℞. Acetum opii f℥ij., acetum lobeliæ, f℥j., aquæ rosæ, ℥ij. Mix. Let half an ounce be inhaled every three or four hours. This is my favorite formula for the use of opium.

No. 10.—℞. Extractum conii, ℥j., spiritus chloroformi, ℥j., spiritus vini rectificati, ℥ij. M. Add two drachms to two of water, and let it be inhaled.

No. 11.—℞. Tinct. stramonii, ℥j., aqua rosæ, ℥v. Mix. Use half an ounce for an inhalation.

Stimulant inhalations are employed in chronic laryngitis, bronchitis, and phthisis, where there is relaxation of the mucous membrane and increased secretion. They not unfrequently relieve the cough and unpleasant sensations, when the narcotics would have no influence. The simpler the agents, the better their influence as a general rule.

No. 12—℞. Vinegar of lobelia, f℥j., comp. tinct. of lavender, f℥ij., infusion of chamomile, f℥v. M. Use half an ounce with the atomizer every four hours.

No. 13.—℞. Tinct. iodine, f℥ss., Rectified spirit, f℥iij. Mix. To be used in the same way. Care should be employed at first not to get the inhalation so strong as to produce irritation, and the vapors breathed through a glass funnel prevent soiling the face and clothes. If found beneficial the strength may be gradually increased.

No. 14.—℞. Tincturæ iodi., min. xxx., aqua calidæ, f℥iv. Mix. To be cautiously inhaled with a common inhaler. (Tanner.)

No. 15.—℞. Olei terebinthiuæ, f℥j., aquæ calidæ, f℥vj. Mix. To be employed in the same way in cases of chronic bronchitis with excessive secretion. (Tanner.)

No. 16.—℞. Creosote, min. xxx, aquæ, ℥iv., tinct. lavendulæ comp., f℥ij. Mix. May be employed in either way.

No. 17. ℞. Nitrate of silver, gr. xx., distilled water, ℥ij. Mix. This should be used with the glass funnel, and its strength carefully proportioned to the susceptibility of the patient.

No. 18.—℞. Permanganate of potash, gr. xv., distilled water, ℥iv. It may be best to commence with a weaker solution than this, but in some cases the strength may be increased to ten grains to the ounce of water.

It will be found very serviceable in cases where there is profuse expectoration.

No. 19.—℞. Solution of chlorinated soda, f℥ij., rose water, f℥iv. This also will have to be employed with care.

No. 20.—Lime water of full strength, with or without the addition of morphine, conium, or other narcotic, is an excellent remedy in cases, where expectoration is difficult.

The bitter tonics are employed when there is debility and relaxation, and very frequently serve as vehicles for the use of the narcotics. The infusions seem to answer a better purpose than the tinctures, but either may be used. I have employed the German chamomile with much advantage, and also the hydrastis and the chiretta. An infusion of Peruvian bark answers an excellent purpose in many cases, and I not unfrequently order the tincture. An infusion of senega and serpentaria may also be used. The solution of persulphate of iron and perchloride of iron is occasionally useful, but generally proves too irritating.

No. 21.—℞. Quiniæ sul. gr. xx., dilute sulphuric acid, f℈ij., acetum opii, f℥ij., infusion of hops, ℥iv. Mix.

Astringents are not very frequently employed, but may be occasionally used with advantage. A solution of tannic acid will be found as useful as most of the vegetable astringents.

No 23.—℞. Tannic acid, ℈j., quinine, gr. x., aromatic sulphuric acid, ℈ij., rose water, ℥iv. Mix. Use half an ounce for an inhalation.

No. 24.—℞. Alum, ℈ij., tincture of cinnamon, ℈ij., water, ℥iv. Mix. To be used as the preceding.

No. 25.—℞. Red oak bark, yellow dock root, aa. ℥j., Make an infusion with two pints of boiling water and use as above.

The following are the inhalations employed at the Consumption Hospital London, as given in Squires' Pharmacopœia's of the London Hospitals. They belong to the class of volatile inhalations, and are employed with the apparatus represented in Figures 1 and 2.

No. 26.—Inhalatio Acid Hydrocyan. Dilute Hydrocyanic Acid 10 to 15 mins.; for one inhalation.

No. 27.—Inhalatio Æther. Chlor. c. Hyoscyam. Chloric Æther 30 mins.; Tincture Henbane 30 mins.; Infusion of Hop, or water 8 oz.

No. 28.—Inhalatio Camphoræ. Spirits of Camphor 1 to 2 drms.; boiling water 8 oz.

No. 29.—Inhalatio Chlorinii. Chlorinated Lime 2 oz.; for one inhalation.

No. 30.—Inhalatio Chloroformi. Chloroform 15 mins.; for one inhalation.

No. 31.—Inhalatio Coniæ. Coniine 1 gr.; Spirits of Wine 10 mins.; water ½ oz.

No. 32.—Inhalatio Creasotum. Creasote 6 mins., water ½ oz.

No. 33.—Inhalatio Iodinii. Iodine 2 grs.; Spirits of Wine 10 min.; water ½ oz.

No. 34.—Inhalatio Lupuli. Hops 1½ oz.; boiling water 20 oz.

No. 35.—Inhalatio Opii. Extract of Opium 3 grs.

In using inhalations we are guided almost entirely by the sensations ot the patient, as regards their frequency, duration, and continuance. If it gives relief from the cough, and the patient breathes freer and easier, it is doing good, but if the breathing becomes more difficult, though the cough may be relieved, it is doing harm.

IV.

INHALATIONS IN SPECIAL DISEASES.

We employ inhalations in affections of all parts of the respiratory mucous membrane, and contrary to the general opinion, they are found more markedly curative in some acute than in chronic diseases. As heretofore remarked, they are principally palliative in chronic disease, and must be associated with appropriate general treatment. It will not be necessary to consider each affection at length, but only as it is influenced by inhalations. These will replace those remedies that are employed for their topical influence, but will in no other respect interfere with the treatment commonly pursued. What is said, therefore, may be regarded in the light of addenda to our works on practice.

With a strong desire to present the subject in such a light that it will be found to stand the test of experiment, it is not improbable that it may occasionally be too highly colored. This cannot be avoided, when a subject is new and experience comparatively limited.

CATARRH.

In confirmed catarrh, inhalations have proven of marked advantage, and in some cases curative. The present winter we have had an extended experience in this disease, and whilst in some cases they seemed of little benefit, in others they surpassed the usual means. Of course, in the early stage of the affection, I should order instead, two grains of Opium at bed-time, or fifteen grains of Bi-carbonate of Ammonia, or a full dose of Tinct. Gelseminum, with rest until noon of the next day. These are means that are speedily curative, and quite certain in their action. Yet when the case has progressed until the mucous membrane is thickened, with abundant secretion of mucus or muco-pus, we do not expect this speedy influence.

In this case I frequently order an inhalation of lime water, every three hours with Seigle's apparatus, especially if the secretion is tenacious and removed with difficulty. An inhalation of vinegar and water, or of an infusion of chamomile, or hops, does very well. Formula No. 9 is an excellent inhalation when there is much fullness with closing of the nose. Formula No. 16 will be found useful in protracted cases when the discharge is abundant.

OZÆNA.

In the treatment of ozæna, inhalations will be found to be but palliative in the most of cases, yet they have proven of sufficient importance to stimulate further investigation. The most efficient means of bringing remedies in contact with the diseased mucous surface, is by the new hydrostatic method, or by the use of the common pump syringe. Any remedy may be thus directly applied, and in such quantity as is necessary.

The use of inhalations of water, water and vinegar, or some of the simple infusions spoken of in the first part of the preceding chapter, answer an excellent purpose in allaying irritation. Lime water, solution of Chlorate of Potash, or solution of Chlorinated Soda diluted, arrest the fetor of the discharges, and act as gentle stimulants. The tonic and astringent inhalations will occasionally be found useful.

CHRONIC PHARYNGITIS.

In chronic pharyngitis, especially if the disease extends above the soft palate, inhalations will be found useful. Occasionally these cases are found very stubborn, and cause much annoyance. The local treatment with nitrate of silver is frequently effectual, but in many cases, the affected parts are not reached, and the disease continues. The use of the universal syringe with the curved perforated tube furnishes a good means to make local applications behind and above the palate, and may be employed with advantage.

In using inhalations, I prefer, if the patient can, that they be drawn into the mouth, and forced out through the nose. If they cannot be used in this way, they should be inhaled partly by the nose and partly by the mouth. The tonic infusions with Chlorate of Potash are the agents most frequently beneficial. Formula No. 23 and 24 will be useful when the muocus membrane is relaxed and flabby.

TONSILLITIS.

In the treatment of acute inflammation of the tonsils. I have found inhalations, with Seigle's apparatus, more speedily beneficial than any other method of treatment. Every physician in active practice knows the almost entire uselessness of the common means employed, and to the sufferer from frequent attacks, any means that would offer a prospect of relief would be gladly welcomed.

I generally order an inhalation of a strong infusion of German Chamomile, or of one part of the tincture to six of water. Tincture of Aconite, (the root), thirty drops to half an ounce of water is also very good. A decoction of hops or tansy may be employed instead of these, or in some cases, Formula 9 or 10. The inhalation should be continued from five to ten minutes, and repeated sufficiently often to give the patient ease, if the case is severe. In milder cases two or three times a day will be found sufficient. If there is much febrile action I order small doses of Aconite, as—℞ Tincture Aconite, rad. ʒj; Aqua, ʒiv: M. Give a teaspoonful every one or two hours.

CYNANCHE MALIGNA.

In this affection, inhalations will be adjuvant to the general treatment, as is all local means. The relief of the stomach by a thorough emetic and the use of quinine and iron, with chlorate of potash or sulphite of soda are the curative agencies employed. Still as the patient suffers greatly from the throat, and has diffi-

5

culty in using a gargle effectually, we will find inhalations of sufficient importance to employ them.

We have here an atonic condition of the mucous
membrane, a feeble circulation and innervation, and arrest of the nutritive processes. Our topical applications must therefore be stimulant, and as far as possible of such character as will arrest the destructive met-
amorphosis so rapidly going on. An inhalation of a
saturated solution of Chlorate of Potash, or Sulphite of
Soda answers an excellent purpose. I have found more
benefit, however, from a strong infusion of Baptisia
Tinctoria in equal parts of vinegar and water, than from
any other means. This is also an excellent gargle.

DIPHTHERIA.

In this affection, the employment of inhalations will
be found a most important addition to the treatment.
So far as local treatment proves beneficial, we can get
speedier and better results from remedies used in this
way than in any other.

Experience has conclusively proven that the local
affection is dependent upon an impairment of the
blood, and a correct treatment must be directed to this.
Still the local manifestation in the throat, not only
causes much suffering, but if allowed to progress, reacts
and increases the general disease. It may also prove
fatal by destroying the vitality of the structures affected.
or by extending to the larynx.

This local affection is not to be arrested by the use of
escharotics, as was first attempted, and many physicians
now only use a solution of Chlorate of Potash as a gargle, or something equally simple and mild.

An inhalation of a saturated solution of Chlorate of Potash (with an atomizer), or a solution of Sesquicarbonate of Ammonia or of Sulphite of Soda, will not only give temporary ease, but materially assists in the cure. The employment of a decoction of tansy, or of German Chamomile, Baptisia Tinctoria, or of simple vinegar and water, vaporized in the common manner, prove useful. Permanganate of Potash in the proportion of grs. ij to grs. xv, to the ounce of water, is a good application. Formula 22 is a very good one, as is No. 19. The importance of these means in cases of children too young to gargle, will readily be seen.

If the disease extends to the larynx, these means are of still greater importance. Many physicians now consider the case nearly if not quite hopeless, but I am satisfied that if inhalations are properly employed, quite a large number of these cases may be saved. Even the employment of the vapor of vinegar, or of equal parts of vinegar and water, is attended with marked benefit. The first case of diphtheritic laryngitis in my practice, occurred in the winter of 1860. The patient, a young lady aged 19, was attacked with diphtheria on Friday evening, and by midnight on Saturday, it seemed almost impossible for her to live, respiration was so difficult. The common means employed had been continuously used, but so far without benefit. Inhalations of the vapor of mild cider vinegar were commenced and cautiously employed, and by Monday evening she was out of danger.

In a second case, Mrs. M——, aged 36 years, the diphtheritic laryngitis commenced on the fourth day of the disease. Its progress was very rapid, and the treatment pursued seemed to give but little relief. Becoming much worse at night, I was sent for, but being away from home, the nurse suggested the use of the vapor of a decoction of tansy. When I reached the house, she had been using it about an hour, with some relief: it was continued, and with the internal use of nauseants, she was breathing freely in twenty-four hours.

These were marked cases, and the action of inhalations were so decidedly beneficial that I continued their use in every case. The infusions first named under the head of relaxants, or under the head of tonics, acidulated with vinegar, will be found most generally available. But I have no doubt that the use of lime water as recommended for pseudo-membranous croup will be found equally useful in this disease.

SCARLATINA.

In scarlatina and anginosa maligna, the disease of the throat sometimes assumes great importance, and demands much care in its treatment. If the patient is of that age that gargles or other local applications can be thoroughly employed, we may get along well enough, but in young children we will find an addition to our therapeutic resources necessary. This addition I think will be found in the judicious use of inhalations.

Those that have given the best results in my practice, are an inhalation of vinegar and water, of an infusion of Baptisia or Chamomile, or a solution of Chlorate of Potash, or Hydrochlorate of Ammonia. Not only does the employment of inhalations relieve the disease of the throat, but it quiets that restlessness and irritability of the little sufferer, that proves so exhaustive, and renders it so much more difficult to manage.

MEASLES.

The irritation of the respiratory mucous membrane, causing the harrassing cough, is the most troublesome part of this disease. It is not only troublesome, but many times is not amenable to the usual cough remedies, and will continue until some serious lesion of the respiratory apparatus ensues. Leaving out the Drosera, which I regard as almost a specific in many of these cases, I prefer to treat this bronchial irritation with inhalations.

Formulas 9 and 12 are very good in these cases. Or we may use either of those from 1 to 6, or the sedative formula from the Consumption Hospital. Not unfrequently an infusion of chamomile, or vinegar and water, if used during the eruption, will be all that is required.

If an irritation of the bronchii and harrassing cough continues afterward, Formulas No. 10 and 11 will be found useful. Or if the secretion is tenacious, and expectoration difficult, they may be alternated with lime water. An infusion of the common red clover taken internally, and used as an inhalation, sometimes gives speedy relief.

PERTUSSIS.

Hooping cough is now treated by most physicians by empirical remedies, or as some would say by *specifics*. The employment of Belladonna, Nitric Acid, Drosera, or Trifolium-in-Fœno, rarely fails of giving the necessary relief, though we may be unable to tell how they act. The last remedy deserves especial mention, as it is simple, common, and very efficient. Take of common red clover dried, a sufficient quantity, cover it with boiling water, and after it has stood two hours, strain with pressure and sweeten. I usually give it in doses of a teaspoonful to a tablespoonful every hour or two.

Inhalations are palliative only, and their influence brief. Still there are some cases in which this influence is desirable. I would place more dependence in Belladonna than any other agent, used as follows—℞ Tincture of Belladonna, ℥ij; Alum, ℥ij; Rose Water, ℥vj: M. Used in cases where there was relaxation of the mucous membrane, and secretion was abundant. If the mucous membrane is dry, with redness and dryness of the throat and fauces, I would change it thus—℞ Tincture of Belladonna, ℥ij; Nitric Acid, gtt. xx; Water, ℥ij. One or two drachms of these may be used at a time, and repeated as often as necessary. Formula No. 11 is a good inhalation also.

CROUP.

This means of treatment is employed with decided advantage in croup, in fact, in some cases, I place much reliance upon it. Spasmodic and the milder form of mucous croup is readily treated by the common means, though even here, the vapor of water, or of water and vinegar will be found of assistance.

For ten years past I have never treated a severe case of mucous or membranous croup, without making inhalations of vapor an important means. It allays the irritation and produces relaxation of the intrinsic muscles of the larynx, and thus lessens the difficulty of breathing. And increasing secretion, it promotes expectoration in the mucous variety, and loosens the pseudo membrane in the other. An infusion of Hops, of Chamomile, or of Tansy, acidulated with vinegar, may be employed instead of water.

In pseudo-membranous croup, a late writer recommended the employment of Lime Water as an inhalation, claiming that it may be regarded almost as a specific in this serious disease. I have had only an opportunity of testing it in one case, but this was a very marked one. Finding that the inhalation gave relief, it was continued without internal treatment other than small doses of Veratrum, and the application of cloths wrung out of hot water externally. I have used it in one case of mucous croup with like advantage. I employed Lime Water of full strength, using it every hour at first, and for fifteen minutes at a time, then every two or three hours. This remedy deserves attention, and should be thoroughly tested.

As illustrative of the general impression among physicians that inhalations may be of advantage in these affections, and the imperfect knowledge on this subject, I will make two or three quotations from the recent work of Dr. Prosser James on "Sore Throat." He says: " The contact of watery vapor with inflamed mucous membrane is very soothing. It cuts short the congestive stage of the disease, by a supply of moisture, and removes the dryness, heat and itching. It penetrates through the respiratory tract, and often produces more calm than powerful anodynes. Moreover, the vapor may be easily medicated. The common inhalers are more trouble than they are worth. I usually get the patient to breathe through a large cup-sponge which has been dipped in hot water and rapidly squeezed. A less efficacious method is to lean over a large basin filled with boiling water. The heat may be kept up by the aid of a spirit lamp."

Speaking of the benefit to be derived from the continuous application of hot water to the throat in croup, he says: " Dr. Graves considered he saved several patients by this method, but that it is only applicable to cases seen at the onset." Nor did he propose to give up other remedies. " I go a step further, and *apply the hot water to the inside of the larynx.* Inhalations have already been spoken of in throat disease. It is obvious that their application in the usual way would be a matter of considerable difficulty in children. Moreover, in such a disease as croup, the inspiration is so impeded as of itself to constitute an objection both to the sponge and the inhaler. It has appeared to me, however, that if the little patient could be kept in a warm and moist atmosphere, great benefit must accrue."

CHRONIC LARYNGITIS

It has been claimed that inhalations are the most important means of treatment in this stubborn affection, and when used properly will effect a cure if this is possible. With such an opinion, I commenced their use, but have learned by experience that they are only adjuvants to other treatment.

In chronic laryngitis, one of the most important indications is to keep the organ quiescent, and a failure in this will entail entire failure, no matter what treatment may be adopted. To the extent then that we can employ inhalations to relieve irritation, and check cough, they become useful in this respect. The sedative inhalations may be employed for the purpose, but we will find Formula 9 and 10 most generally useful.

When the disease has progressed to change of structure and ulceration, the inhalations of Iodine or Nitrate of Silver, as in Formula 14 and 17, may be employed with advantage. But in a majority of cases, the tonic infusions, with Opium, Morphia, or Stramonium will be most generally useful.

Permanganate of Potash, gr. iij to grs. x ; to water, ʒij, has been employed with benefit , when the secretion was muco-purulent.

We have the same guide to their use here that we have in phthisis. If the irritation and cough are relieved, the patient breathing more freely, the inhalation is of advantage, but if, with the arrest of cough, there is a feeling of oppression, they should be suspended.

Dr. Copeland says : " The inspiration of dry or moist vapor has been recommended in *phthisis laryngea* and in other affections of the respiratory apparatus : but those which have been employed, and often too empirically prescribed, have been either too acrid, stimulating or concentrated : and not being confined in their operation to the larynx, but acting on the respiratory surfaces generally, have proved more injurious than beneficial. The action of these cannot be limited : and hence, only those which I have advised above, and which are balsamic, aromatic, emollient, and narcotic, and cannot injure the lungs, should be employed." I think this will be the experience of every one that has employed inhalations to any extent; at least it has been mine.

M. Trousseau, speaking of the inhalation of pulverized fluids, says : " As to the therapeutical effects of pulverized inhalations, in no class of cases is it more

apparent than in the granular condition of the mucous
membrane of the pharynx and larynx, termed ' dys-
phonia clerincorum,' and so common in preachers,
orators, singers and those who habitually overexert
their vocal powers. About a year ago, a woman affected with small pox came into my ward in the Hotel
Dieu. She was suddenly seized with œdema of the
glottis, of so rapid a character as to immediately endan-
ger life. After requesting M. Robert to be in readiness
to perform tracheotomy, I resolved to try the effects of
inhalation, and caused her to breathe a pulverized solution of tannin. So rapid and complete was the relief,
that by the evening, all danger had vanished, and the
operation was dispensed with. Quite recently, I met
with a case of the same affection, œdema glottidis, in a
phthisical patient, who was pregnant and near her time.
By the use of the pulverized inhaler, I was able to pro-
long this woman's life until after the birth of the child,
and attain a result I should otherwise have despaired of
accomplishing. In syphilitic affections of the larynx,
I have been equally happy, and very willingly bear testimony to the efficacy of the invention."

Dr. Maddock reports two cases of chronic laryngitis
in his work, in which he attributes the cure principally
to inhalations. In one, he employed Belladonna, with
the effect of arresting the irritation and cough, and all
symptoms of local disease in seven weeks. In the second, he used inhalations of Iodine and Conium, with
like success.

ACUTE BRONCHITIS.

It has long been noticed that it was essential to have a moist, warm atmosphere in the treatment of this affection, and quite a number of writers place great stress upon it. It is true, their advice is not followed once in a hundred cases; indeed most physicians think they have done their entire duty to their patients when they order their expectorant mixture.

The best observers go further than this, and recommend the use of demulcent and mild narcotic infusions, vaporized for inhalation. Thus Dr. Copland remarks: " The *inhalation of emollient and medicated vapors* is occasionally of much benefit in the sthenic form of the disease, but chiefly in its first and second stages. The vapor arising from a decoction of Marsh Mallows, or from Linseed tea, or from simple warm water, is the best suited to this state; and should be employed from time to time, the *temperature of the apartment* being duly regulated through the treatment, and constantly preserved from about 66° of Fah., to 75°."

And further on he says: " When the expectoration becomes whitish, opaque and thick, the vapor may be rendered more resolvent, by adding a solution of Camphor in vinegar, and extract of Conium or Hyoscyamus to the hot water, or to the emollient infusions now mentioned; and in the asthenic variety, particularly when the difficulty of expectoration and the fits of dyspnœa are distressing, or when the excretion of morbid matter is impeded or suppressed for want of power the medicated vapors and gases recommended in the chronic state of the disease may be tried."

In the first stage of the disease, inhalations of the vapor of water and other mild agents, give more relief than any other means I have employed. They are pleasant to the patient, relieve the dryness and sense of irritation, check the harrassing cough, and speedily reestablish secretion. Thus they accomplish that for which *nauseants* are usually employed, and in one-half or one fourth of the time, and without the decidedly unpleasant symptoms that attend the action of that class of remedies. An inhalation of the vapor of water sometimes answers every purpose. It may be used as often as necessary, by any of the methods named, or by the employment of Seigle's apparatus, which I think decidedly the best.

In some cases, an infusion of Hops, or of Poppy heads, seems to act better than the water alone. An infusion of Chamomile flowers, or a small portion of the tincture added to water will also prove useful. In some cases I have used Tincture Aconite, f℥j to water f℥iv. If the circulation is somewhat feeble, I prefer Belladonna, or use it in conjunction with aconite in the proportion of f℥ss to f℥j, to water f℥iv. In some cases the patient suffers greatly from difficult breathing and oppression, the result of contraction of the muscular fibres of the bronchii. Here I use Stramonium in the proportion of—℞ Tincture of Stramonium, f℥ss; water, f℥iv.

If the patient's tongue is red, dry and slick, we will find it of advantage to acidulate the inhalation either by the addition of vinegar or Hydrochloric Acid. I use the last in the proportion of Dilute Hydrochloric Acid, ℥ss to water, or one of the infusions named, f℥iv. But if the tongue is broad, white and pasty, I should employ Lime Water, of full strength, or diluted with one part of water. With these inhalation, I employ the direct sedatives, as—℞ Tincture of Veratrum, f℥j;

Tincture of Aconite, gtt. xx; Water f℥iv : M. Give a teaspoonful every hour until the fever abates, using the sponge bath and hot foot bath, and afterward every two or three hours. As a local application to the chest I prefer the mush poultice.

It is very rarely that other means than this will be needed, and the treatment is pleasant and the cure speedy. But if the disease progresses, and the expectoration becomes abundant and muco-purulent, we may then employ the stimulant inhalations heretofore mentioned,—the Acetous Tinctures of Lobelia and of Opium, Formula 9 and 12, or of Chlorinated Soda, Formula 19.

ASTHENIC BRONCHITIS.

In this affection we can obtain more decided results from remedies employed in inhalation, than in any other way. Of course we would not withhold Quinine, Iron and stimulants, as these are important parts of the treatment. So little are inhalations thought of, I have not unfrequently been asked, whether they were better than a tonic and stimulant course of treatment in such affections. It must be understood, in order to get a fair idea of their value, that they do not replace other remedies, but are always to be regarded as additional means. Or if they replace any agents, it is that indirect class termed expectorants.

In this disease, inhalations of a stimulant character are most generally employed, though some others will be found equally useful. Thus I have obtained most marked benefit from the use of Belladonna and Stramonium, as in the following formula—℞ Tincture of

Belladonna, f3ij; Dilute Acetic Acid, f3ij; water, f3iv; Mix.—℞ Tincture of Stramonium, f3ss; Dilute Acetic Acid, f3ij; water f3iv: Mix. I order an inhalation of from two drachms to half an ounce, every two hours with an *atomizer*, or vaporized in the common manner. If it produces headache or dizziness, lessen the strength of the solution. An inhalation of the Tincture of Hamamelis, f3ss to water f3iv, answers a good purpose.

In some cases we find Formula 9 and 12 very good remedies, the Lobelia and Opium acting as powerful stimulants to the relaxed mucous membrane. The turpentine inhalation, Formula No. 15, answers a very good purpose when secretion is profuse, as will the solution of Permanganate of potash, Formula No. 18, or Chlorinated Soda, No. 19. In other cases the astringents may be employed, as of Tannic Acid in Formula No. 23, or Alum in Formula 24. Or the resins and balsamic agents may be used as named under the head of chronic bronchitis.

PULMONARY APOPLEXY.

We use this name for want of a better, as it is generally understood to express great congestion of the structure of the lungs, with effusion into its parenchyma, into the air-cells and bronchial tubes. In some cases there is hemorrhage, but in more a profuse secretion of a rusty, brownish, or dirty sputa.

These are rare cases, but very difficult to manage, and not unfrequently prove fatal under the ordinary treatment. I adopt the following plan. Give internally,—℞. Tincture of Lobelia, Compound Tincture of Lavender, aa f3j; Simple Sirup, f3ij: Mix. Give in doses of one teaspoonful every half hour at first, then every one and two hours. Use an inhalation of Cider Vinegar, at first very slowly, increasing the quantity of the vapor, as the patient can bear it. In using Seigle's apparatus, I order—℞. Tincture of Stramonium, f℥j; Dilute Acetic Acid, f℥vj: Mix. This treatment has proven so satisfactory, that I feel that I can hardly recommend it too strongly. Within the last week I have had such a case, the symptoms being so severe within six hours, that no person who saw him supposed he could live until night. Expectoration was so abundant, that it required every effort to free the tubes from the rusty sanious material. Yet this man was relieved in twelve hours, and was up on the fourth day.

CHRONIC BRONCHITIS.

Inhalations have been used to a greater extent in this disease than in most others, at least we find more written upon it. As the disease is one involving change of structure and function of a part readily reached by medicated vapors, we can easily see why it will be amenable to this plan of treatment. Still, as in preceding cases, we must not neglect the equally if not more important object, of improving the general health by the judicious use of bitter tonics, Iron, Hypophosphites, Cod-liver Oil, etc. All physicians who are successful in the treatment of chronic disease, especially diseases of the respiratory organs, recognize this fact, *that the local affection is the more readily cured, the better the general health is.* To this rule there are no exceptions, hence the importance we attach to a general tonic and restorative treatment.

Dr. Copland seems to have had a very clear conception of what might be accomplished by the employment of inhalations in this affection, and employed them to some advantage; but from the imperfection of the apparatus, they were not used to any considerable extent. It will be instructive to read what this author says, upon the subject of inhalations.

"Notwithstanding the unsuccessful attempts of Beddoes to revive the practice, by employing the elementary and permanently elastic gases, but according to views too exclusively chemical, the practice of inhalation has long been neglected or undeservedly fallen into the hands of empirics. Very recently, however, it has been brought again into notice by M. Gannal, Mr. Murray, and Sir C. Scudamore; and *Chlorine Gas*, and fumes of *Iodine*, and watery vapor holding in solution various *narcotics*, have been recommended to be inhaled. I have tried those substances in a few cases of chronic bronchitis; but in not more than two or three cases of tubercular phthisis. The Chlorine was used in so diluted a state as not to excite irritation or cough. The Sulphuret of Iodine, and the *Liquor Potassii Iodidi concentratus* were also employed; one or two drachms of the latter being added to about a pint of water, at the temperature of 130°, and the fumes inhaled for ten or twelve minutes, twice or thrice daily. The tinctures or extracts of Hyoscyamus and Conium, with Camphor, added to water at about the above temperature, were likewise made trial of; and although the cases have been few in which these substances have been thus used by me, yet sufficient evidence of advantage has been furnished to warrant the recommendation of them in this state of the disease.

" *Inhalations*, also, of the fumes of the *balsams*, of the

terebinthinates, of the odoriferous *resins,* &c., are evidently, from what I have seen of their effects, of much
service in the chronic forms of bronchitis : and I believe that they have fallen into disuse, from having
been inhaled as they arise in a column or current from
the substances yielding them, and before they have been
sufficiently diffused in the air. When thus employed,
they not only occasion too great excitement of the
bronchial surface, but also intercept an equal portion of
respirable air, and thereby interfere with the already
sufficiently impeded function of respiration. M. Nys-
ten has shown (*Dict. des. Scien. Med.* t. xvii. p. 143,)
that ammoniacal and other stimulating fumes, when
inhaled into the lungs in too concentrated a state, produce most acute inflammation of the air-tubes, generally terminating in death ; and has referred to a case in
which he observed this result from an incautious trial of
this practice. I conceive, therefore, that the vapors
emitted by the more fluid Balsams, Terebinthinates, the
Resins, Camphor, Vinegar, etc., and from Chlorine and
the preparations of Iodine, should be more diluted by
admixture with the atmosphere, previously to being
inhaled, than they usually are. According to this view,
I have directed them to be diffused in the air of the
patient's apartment, regulating the quantity of the
fumes, the continuance of the process, and the frequency of its repetition, by the effects produced on the
cough, on the quantity and state of the sputa, and on
the respiration. The objects had in view have been
gradually to diminish the quantity of the sputum, by
changing the action of the vessels secreting it; without exciting cough, or increasing the tightness of the
chest, or otherwise disordering respiration. From this
it will appear, that the prolonged respiration of air
6·

containing a weak dose of medicated fumes or vapors, is to be preferred to a short inhalation of them in their more concentrated states. The want of success which Dr. Hastings and others have experienced, evidently has been partly owing to the mode of administering them, and partly to having prescribed them inappropriately. When the patient complains of acute pain in any part of the chest, as in some of Dr. Hasting's cases, they are as likely to be mischievous as beneficial. Where benefit has been obtained, it will be found that it was when the fumes of the more stimulating of those substances were diffused, in moderate quantity, in the air of the patient's apartments; or when he passed, at several periods daily, some time in a room moderately charged with the vapor or fumes of the substance or substances selected for use."

In my practice I very rarely employ the sedative inhalations first named, unless there has been great irritation; in such cases the Formulas from 1 to 6 may be used, or the sedative Formula of the Consumptive Hospital, No's 26 and 27. Using Seigle's Apparatus, I employ an infusion of Lobelia Herb or Verbascum, or Hops, if there is dryness of the mucous membrane. In some cases we may alternate these with Formula No. 9, or No. 11, either of which will prove useful.

If the secretion is tenacious and raised with difficulty, causing hard and prolonged coughing, I employ an inhalation of Lime Water with Stramonium, Belladonna, or Morphia, as may seem indicated by the particular case. Occasionally we will find the Belladonna and Stramonium to act better, as in the Formulas under the head of Asthenic Bronchitis.

If expectoration is free but not excessive, with a feeling of oppression and debility, as regards the respira-

tory function, we will find the Belladonna, alternated
with the Chlorine, as in the Formulas No. 19 and 29,
very good. Iodine, very dilute, may be inhaled with
Conium or Belladonna. But in some of these cases,
Opium will give the best results, as in Formula No. 9,
or associated with some tonic infusion.

Where secretion is profuse, we may employ the stimulant inhalations, or the astringents. I have obtained
very marked benefit from the. use of Iodine, and from
the Permanganate of Potash, and in these cases both
Iodine and Chlorine are highly recommended by those
who have made most use of them, Chlorine Water,
used with Seigle's Apparatus, would likely be the best
form. Of the astringent inhalations, Formulas No. 23
and 24, may be used with advantage, or the Acetous
Tincture of Opium may be added to either. The ter-
ebinthinate inhalations are used with advantage in these
cases, and occasionally we derive benefit from the Gum
Resins, or Balsams.

ASTHMA.

As a palliative, inhalations will be found more generally available than other methods of using medicine,
being prompt in their action, and much pleasanter for
the patient. Even an inhalation of the vapor of water
gives relief in many cases. I have employed the Tincture of Stramonium in the proportion of f℥ss. to water
f℥iv: with very marked advantage. Belladonna may
be used in those cases in which there is relaxation of
the bronchial tubes. I usually order it in the following
proportion—℞. Tincture of Belladonna, f℥j. to f℥ij;

Water, f℥iv : Mix. Use half an ounce for an inhalation.

In some cases I have used an infusion of Lobelia, and in others of Verbascum, or the Tinctures may be added to water and employed in the same way.

The use of inhalations of Chloroform and Æther as palliatives are well known, and practiced to a considerable extent. They may be associated with the means just named. The sedative inhalations from Nos. 1 to 6 may also be used occasionally.

In those cases where the tongue is broad, pale, and pasty, I have obtained good results from alkaline inhalations. The Lime Water, heretofore named, answers an excellent purpose, or we may use the Chlorinated Soda, Formula 19. Occasionally the Permanganate of Potash, will be found useful, as in Formula 18, this is especially useful when secretion is free.

When the secretion is abundant, and the bronchial tubes relaxed, as in that variety termed *Humoral Asthma*, the remedies recommended for Asthenic Bronchitis will be available.

In describing the treatment of Asthma, Dr. Copland thus writes of inhalations :

"Next and, perhaps, equal to smoking, is the inhalation of simply emollient or of medicated vapors into the lungs. This method of treatment was recommended by Cælius, Aurelianus, Alberti, Mudge, Beddoes, Thilenius, Zallony, Hufeland, Crichton, Forbes, Gannal, Scudamore, and Murray. It is chiefly indicated during the paroxysm, or shortly before its accession. The vapors arising from pouring boiling water upon Camphor, any one of the Narcotic Extracts or Tinctures, or the balsams, are of great advantage when properly

managed. Thus the vapor from a pint of boiling
water poured upon half an ounce of balsam of tolu;
or that from a Solution of Camphor, Balsam of Tolu,
and Extract of Lettuce, or of Conium, in Sulphuric
Æther; or the fumes proceeding from Camphor, Hyoscyamus, and Aromatic Vinegar, mixed together, and
quickened by the addition of some boiling water, may
be employed. A Solution of Balsam of Tolu in Sulphuric Æther, the vapor of boiling Tar diffused in the
air of the patient's chamber, Chlorine Gas much diluted
with common air, and various other medicated vapors
may be tried; but these act chiefly by removing the
viscid phlegm which collects in the Bronchi, and by
exciting the extreme exhaling vessels. I have prescribed the vapor of the *Sulphuret of Iodine* in two
cases: in one of Spasmodic Asthma, with no benefit;
and in one of Humoral Asthma, with only temporary
advantage. Sir C. Scudamore recommends this Formula for the inhalation of Iodine—Ŗ. Iodinii gr. viij;
Potassii Iodidi gr. v; Alcoholis ℥ss; Aquæ Destil. ℥vss:
M. Fiat Mistura. To this he adds Tincture of Conium.
But his directions as to quantity and mode of inhalation are, notwithstanding several attempts to unravel
them, perfectly beyond my powers. I believe, however, that portions only of the above mixture should be
employed for each inhalation. But the observing practitioner will generally be able to apportion the quantity,
as well as to direct the particular materials, for inhalation, according to the peculiarities of the base; bearing
in recollection that the combination of narcotic and
anodyne vapors with the volatile fumes and gases will
generally be of more service in Asthma than the use
of individual substances belonging to one only of these
classes of medicines; and that the more irritating sub-

stances of this description, such as Iodine, Chlorine, and Tar vapor; should be ventured upon only in a very weak or diluted state."

PNEUMONIA.

The opinions of the profession in regard to the treatment of inflammation of the lungs, has undergone a decided revolution in the last twenty years. The old fashioned antiphlogistic treatment has been proven worse than useless, and those *heroic* means, blood-letting, mercurials, and antimony, with their associates, are buried so deep that we trust there will never be a resurrection for them. In theory we may talk about a change of type in disease to account for the change of treatment, and cover the shortcomings of the old practice. But admitting that in olden time diseases were of a more sthenic character, it was only that much better for the patient, as they were more readily managed by simple means.

So entirely futile were the old means, that many physicians, and some of the best writers in England, almost or quite discard medicine and rely exclusively on the mush poultice, and diet and rest.

I would not go this far. Thus I find that in the proportion that I control the circulation by the use of the special sedatives, I control the inflammatory process. And that this is aided by such means as relieve irritation of the bronchial mucous membrane. Thus I order Veratrum in the usual doses, with a mush poultice to the chest. In place of giving the usual expectorant remedies for the relief of the cough, I think I obtain better results from the use of inhalations.

At first I employ the vapor of water, or of an infusion of Hops or Chamomile, as recommended in acute bronchitis. When secretion is established, I acidulate these infusions with cider vinegar. These generally are all that is required, but if the case proves stubborn, and the secretion profuse, or the stage of grey hepatization is reached, I prefer the solution of Permanganate of Potash, or of Chlorinated Soda. It is hardly worth while to name other remedies, as what was stated in the consideration of acute bronchitis is equally applicable here.

PHTHISIS PULMONALIS.

Consumption is now regarded by the better informed physician as a disease of nutrition, and medication is directed to this end rather than to the lungs. Occuring in persons of feeble vitality, either congenital or acquired, we observe a still further depression previous to, or with the development of the symptoms of phthisis. Two causes may be said to exist at the same time, an imperfect elaboration of the blood, and defective secretion. With imperfect or feeble vitality, the blood is always of lower organization than in persons in robust health, and there is always more or less imperfectly formed material that must be removed by way of the excretory organs. We will therefore find them doing a larger amount of labor than in others. In addition, the tissues do not possess so high a grade of vitality and give way much easier. If, therefore, there is such further impairment of vitality in these cases as to lower the organization of the blood, we shall have the material of tubercle (imperfectly formed albumen),

in excess.   If the excretory organs are active, this material is removed by way of the skin, kidneys and bowels, and no deposit occurs.   But should they fail, the
material circulates in the blood as a foreign body, and
will be thrown out as tubercle, whenever an irritation
with determination of blood is set up.

Not only so, but it is clearly proven that the broken
down elements of the tissues is a source of tuberculous
material, in persons of feeble vitality.   If this be the
case, and it is conclusive to my mind, we cannot but
see the imperative necessity of keeping the secretions
free for its removal.

The indications of treatment are therefore three in
number.

1. To place the stomach and digestive canal in good
condition, give the patient an appetite and power of
digestion, and such restoratives as serve to make a better quality of blood.

2. To restore and keep active, secretion from the skin,
kidneys and bowels, and thus remove material that
cannot be used for nutrition of texture, but will be
deposited as tubercle if not removed in this way.

3. To check irritation of the lungs, controling the
cough and determination of blood, and thereby check-
ing the deposit therein until the material may be removed in other ways.

It seems to me that these indications are very clearly
deducible from the pathology of the disease, and experience has proven to me, that the practice based upon
them is quite successful.   Indeed, if they are fully ap-
preciated, it is hardly worth while to speak of special
medicines, as the proper ones will immediately suggest
themselves.   We have, unfortunately, no specifics for

consumption, and as far as treatment proves useful, it must be based upon the principles above laid down.

One great stumbling block in the way of correct treatment, is the fear of increasing the debility by the use of means to establish secretion, and place the stomach and bowels in good condition. It is true, that all debilitating treatment should be avoided, but with the proper remedies to fulfill these indications no debility will result. Thus I do not hesitate to give a thorough emetic, where the stomach is in an atonic condition, and there are morbid accumulations and increased secretion of mucus, and repeat it if necessary. Or, to give a mild cathartic of Podophyllin and Leptandrin, with Hydrastin and Quinine, to get normal activity of the upper intestine, and promote secretion. Stimulate the skin to action by baths, alkaline, tonic, stimulant, astringent, oleaginous, or vapor, as the case may need.

In some cases I have trusted to a great extent to Iron, frequently using the muriated tincture, one part to three of Glycerine, in doses of a teaspoonful four times a day. The Collinsonia Canadensis, is an excellent remedy, fulfilling the double indication of a tonic, and a remedy to control irritation of the lungs and cough. The Hypophosphites will be found useful where there is much excitation of the nervous system, and where there is a languid circulation of blood. The compound syrup is a good preparation, or we may use the Hypophosphites of Lime and Soda in alternation. In place of these, the Tincture of Phosphorus, ʒj, to Simple Syrup, ʒiv, may be given in teaspoonful doses every four hours. Cod-liver oil proves useful when it is readily taken and digested, but if it nauseates and impairs the appetite, it should be omitted. Small quantities of Bourbon Whisky are occasionally used with advantage; but should

7

never be employed as a substitute for food, or to such an extent as to impair the appetite.

In the fulfillment of the third indication we have the use of inhalations in the treatment of this disease. And here I wish to be distinctly understood as not claiming that they will supplant the use of the ordinary means in all cases, but only that they may be used in addition to them, and in some cases entirely replace them.

The agents employed in inhalation will depend to a considerable extent upon the amount and kind of secretion from the bronchial tubes, and the ease and difficulty with which it is raised. Again, we have to take into consideration the state of the mucous membrane, as regards tonicity and circulation, and lastly the general health of the patient.

It may be laid down as a general rule, to which there are few if any exceptions, that an inhalation which relieves the cough, and gives ease and freedom in respiration, cannot but prove beneficial. And, on the contrary, any remedy that unduly excites, causes determination of blood, or gives rise to a feeling of fulness and oppression in the chest, will always prove detrimental. Bearing this in mind we will rarely cause harm by the use of these means, as I have not unfrequently witnessed in the treatment of those who make inhalation a specialty.

If there is dryness of the air passages, I order an inhalation of the vapor of water, slightly acidulated or rendered alkaline, according to the condition of the tongue. If the tongue is red, and inclined to dryness, sometimes slick, I always acidulate the inhalation, either with vinegar, or some other of the vegetable acids. If the tongue is pale, broad, and covered with a pasty coat, I always render them alkaline by the addition of soda or of lime. I attach great importance to these suggestions, as I believe them founded upon an unchanging therapeutic law, which is equally applicable to the administration of remedies by mouth.

An infusion of Hops, of Chamomile, of Marsh Mallow, or even of some of the simple bitter tonics will be found useful here. If there is great irritation and harrassing cough, and these means do not seem to relieve it, the sedative or narcotic inhalations may be alternated with them, or added in proper proportion to them.

Of the sedative inhalations, I prefer Formulas No. 3, 5, 6 and 27, as they allay irritation and cough, without producing oppression. These inhalations will be applicable to those cases in which secretion is moderate, and expectoration not difficult.

If the sputa is tenacious, and raised with difficulty, we obtain the best results from an infusion of Lobelia alone, or with Opium as in Formula No. 9. If there is difficult breathing from contraction of the bronchii, the Stramonium as in Formula No. 11, will prove better. In place of these we may not unfrequently obtain better results from Lime Water, with the addition of a solution of Acetate of Morphine.

If the secretion is abundant, we employ stimulants and astringents. It is in these cases that we obtain benefit from the use of Iodine and Chlorine. We may employ Iodine as in Formulas 33 and 14, with the apparatus represented in Figures 1 and 2. Or with the atomizer as in Formula 13.

In using Iodine, we commence with small quantities, and I prefer it with the vapor of water, as in using it with the atomizer. Occasionally we prescribe Belladonna or Stramonium with it, or Acetate of Morphia, or other of the sedatives.

The treatment of pulmonary consumption by Iodine is very frequent in Belgium, and has been especially recommended by M. Chartroule. Under his directions twenty-eight patients in the hospital were treated by the inhalation of the vapor of pure Iodine, and of this number only eleven could be said to derive no benefit from the treatment. In these unsuccessful cases, the pulmonary lesions were not modified, but still the symptoms were not aggravated in any case. In opposition to the statement that Iodine vapor produced hemoptysis, it was found that pulmonary hemorrhage ceased more rapidly under this kind of treatment than under other plans which are more generally employed. Seventeen patients derived positive benefit from the Iodine treatment, and this improvement was observed not only in relation to the general symptoms, but also to the pulmonary lesion itself, as was proved by percussion and auscultation. Out of the seventeen patients, four might be considered as actually cured. One of the cases of cure is the following:—A youth, sixteen years of age, entered the hospital in such an alarming condition that at first the physicians hesitated to submit him to the Iodine inhalations. He was in a state of great emaciation, and his skin was almost constantly covered with profuse perspiration; he had diarrhœa, which had lasted for two months, and he had repeatedly suffered from hemoptysis. There were very extensive indurations in the lungs, and at the apex of the right lung there was a cavity of some size, as was shown by very obvious gargouillement. The expectoration was also characteristic. After resting a few days, this young man was subjected to the Iodine inhalations, and all the symptoms which had appeared so serious were soon modified in a most remarkable manner. The general symptoms disappeared first, and the body recovered its plumpness with great rapidity. The perspiration, diarrhœa, fever, cough, and expectoration were soon relieved or removed, and six weeks after admission into the hospital the patient went out quite well.

I prefer Chlorine in the form of Chlorine Water, used with the atomizer. That is best, that is prepared by passing the Chlorine disengaged by the action of Hydrochloric Acid and Binoxide of Manganese, into water to saturation. The Middlesex Formula gives a very good solution of Chlorine, and as it is easily prepared, it will be most generally used—℞. Potasssæ Chloritis, ℨij ; Acidi Hydrochlorici, Aquæ Distillatæ aa. f℥ij : Misce. This should be kept in a stoppered bottle in a dark place. In using it from, f℥ss to f℥j may be added to f℥j, of water. In some cases it will have to be used much weaker than this.

Turpentine may be employed with advantage in some cases, either alone, or in connection with the sedatives. Tar water has likewise had considerable reputation, and a quite celebrated physician was accustomed to use rum that had been boiled with tar.

In some cases of phthisis, coal oil has been employed with reported advantage as an inhalation. It seems better adapted to those cases in which there is abundant expectoration, with great relaxation of the bronchial mucous membrane. Add to four ounces of Coal Oil, ten grains of Sulphate of Morphia, and use from two to four drachms as an inhalation.

In place of these we may use the astringent inhalations, but as a general rule they will be found to cause oppression and difficult breathing, and will have to be abandoned.

In concluding this subject, I will make an extract from the work of Dr. Maddock, not that it throws any more light upon the subject, but as showing that authority is not wanting for many of the statements I have here made:

" Dr. Bodtcher, of Copenhagen, has published some interesting observations on the efficacy of camphorated vapors in complaints of the air passages; Raspail has also strongly recommended the use of Camphor in nervous and spasmodic affections of the air passages, the patient taking it powdered, as snuff, or respiring its vapor—small pieces of Camphor being enclosed in a straw, or quill, the ends closed with cotton-wool, the tube then placed in the mouth, and the breath drawn through it. Dr. Harwood speaks highly of the inhalations of Ammonia and Camphor, employed with a temperature of about 100°, which are frequently very useful in relieving distressing symptoms, and in promoting the cure of some affections of the fauces, the larynx, and trachea, among which the most common are hoarseness and loss of voice; their benefit arising either from acting directly on the part affected, or from communicating their influence along a limited extent of certain nerves of the throat, by a sympathetic action. On other occasions, by suitably diminishing the stimulus of these inhalations their influence may be safely extended to more remote parts of the pulmonary nervous system; and hence in some chronic complaints of the chest, in elderly persons, much benefit has attended the addition of a little Ammonia to Ammoniacum, or other expectorants, with a view to arouse and augment nervous power in the lungs, in consequence of its being so far diminished, as to render the removal of phlegm from the air passages very difficult.

"Sir A. Crichton, in an 'Account of some Experiments made with the Vapor of Boiling Tar, in the Cure of Pulmonary Consumption,' after detailing various cases which had come under his treatment, makes the following remarks: 'It must be evident that the Tar fumigation, though most completely successful in some of them, did not produce the same good effects in all; but, on the other hand, the very great relief which every patient experienced at first from it, particularly in the diminution of cough, expectoration and fever, is a fact which ought to encourage us to multiply the trials of this remedy as far as possible. * * * The Tar vapor seems to have healed the ulcers, and removed the inflammation of the tubercles, in the greater number of cases, but I do not believe it produces the absorption of the tubercles themselves. * * * At that period when the cough, expectoration, and hectic fever are greatly subdued by the influence of the Tar fumigation, it seems to me often injudicious to continue it longer, or at least, in so strong a degree as before. Notwithstanding the great power of this means of cure, I never employed it quite alone, but at the same time prescribed internal remedies, such as the nature and urgency of the symptoms seemed to require; but these have been the same as every practical physician has recommended in similar cases.'

"Since the introduction of Tar, by Sir A. Crichton, in the year 1817, numerous trials of it have been made by Lazareto, Hufeland, Dr. Morton, of Philadelphia, Dr. Neumann of Berlin, and many others, all of whom are sceptical of its value as a *curative* agent in pulmonic diseases, but believe it capable of diminishing the night sweats, the expectoration, and hectic fever.

"The mode of using the Tar preparation consists in boiling some common Tar, and adding to each pound from one to two ounces of the Carbonate of Potash, to destroy the Empyreumatic Acid; a small quantity of this is put over a spirit lamp, and by thus disengaging the volatile part of the Tar, which consists of an invisible vapor, the air of the apartment soon becomes impregnated.

"Inhalations of creosote, a preparation of Tar, were extensively tried by Dr. Elliotson a few years ago, but were speedily discontinued as being of little or no value.

"We have employed Tar, Creosote, and that species of Tar termed naphtha, or pyro-acetic spirit, in a great number of cases of pulmonary consumption with the utmost caution and perseverance, and in exact accordance with the directions enforced by their respective advocates as to quantity and quality, but without deriving the beneficial effects which have been attributed to them, and we believe their efficacy is of very limited applicability, and refers only to some few phenomena and effects of the disease.

"Dr. Bennet, in his work entitled Theatrum Tabidorum (chap. De Halituum et Suffituum), records several cases of pulmonary disease successfully treated by inhalations of various gases and watery vapors, with accounts of the fumigating apparatus, and recipes for the remedies.

"Dr. Cottereau, of Paris, has communicated, in the Journal Hebdomadaire, and the Arch. Gen. de Medecine, for 1830, several highly important cases of pulmonary consumption, in which perfect recovery ensued from the use of chlorine inhalations. Mr. John Murray, in a most interesting and able work on pulmonary consumption, has also narrated numerous cases of pulmonic disease which had been cured by the same remedy. Dr. Elliotson, in the Lancet, No. 402, observes in his admirable lectures, that he has seen many cases of tuberculous consumption and diseases of the air passages in which the more distressing symptoms were quickly relieved by the inhalation of chlorine; but hesitates to give a decided opinion of its curative effects until he has made further trials. Dr. Elliotson at the same time remarks, that the medical profession have been much to blame for neglecting the inhalation of various substances, and allowing their patients to die under the old 'jog-trot' system, well established as unsuccessful; and that the duty they owe to themselves and their patients demands that they should not persist in affording alleviations only, when there was the slightest possibility of accomplishing more good than before by any new means. Dr. Elliotson adds, that 'it shows a very narrow mind to set one's face against attempts at improvement; and I, therefore, give credit to all my medical brethren who suggest anything new, and still more to those who make exertion to carry such things into effect.' "

V.

ON THE USE OF ATOMIZED FLUIDS IN OTHER AFFECTIONS.

Though not entirely pertinent to the subject of the work, I cannot let the opportunity pass of saying something with regard to the use of atomized fluids in other diseases, than those of the respiratory organs. What I have to say here, is the result of my own experience, and will have to be corrected and enlarged by the practice of others. I am satisfied, however, that the facts will stand the proof of experiment, and prove a valuable addition to our therapeutic resources.

Whilst we employ the same remedies that we would use as local applications in the same diseases, we obtain a far greater influence by their minute subdivision, and by their equal application to the diseased surface. In many cases, a local application, however carefully applied, only reaches a part of the affected surface, and is very speedily removed by the natural secretion of the part. With the atomized fluid, it penetrates every part, and is continually renewed as long as the operator desires, whether it be five minutes or one hour.

THE EYE.

The applicability of this means of medication in diseases of the eyes will be readily seen. The warmth and moisture relieves for the time that irritability that so frequently prevents the proper application of eye-washes. The fluid reaches every part, is equally distributed, and its application continued as long as desired. If these are facts, and any person may readily test them for himself, it will be seen that this method is a distinct step in advance of the common practice.

If they were no better in the case of the adult, this could not be claimed in treating diseases of the eyes in children. Here we find it almost impossible to use even the most simple collyria with good results, and even when it is applied, the force used, and the resistance of the child, does more harm than the local application does good. But with Seigle's atomizer we have no difficulty whatever, as the soothing influence of the vapor more than counterbalances the effects of the medicine.

ACUTE CONJUNCTIVITIS.—This is the case in which atomized fluids will be found most strikingly beneficial, as the influence of the vapor is emollient and soothing, and relieves of itself, to some extent, the irritation upon which the determination of blood depends. We may employ the vapor of water alone, or we may use any mucilaginous infusion. I generally order—℞ Tincture of Belladonna, f℥j; water, f℥iv : or—℞ Tincture of Belladonna, f℥j ; Tincture of Gelseminum, f℥ij ; water, ℥iv : or—℞ Aromatic Sulphuric Acid, gtt. xv ; Morphia Sul. gr. ij; Aqua, f℥ij. In place of the Tincture of Belladonna, Atropia might be used, but it would have to be very largely diluted.

PURULENT CONJUNCTIVITIS.—In purulent conjunctivitis these same remedies may be employed, the combination of Aromatic Sulphuric Acid and Morphia being
especially useful. In other cases we may use an infusion of Hydrastis or Baptisia, or of Chamomile, alternated with the common collyria. A weak solution of
Carbonate of Ammonia, or Chlorate of Potash, say
from one to five grains to the ounce of water, will prove
useful. Or if the discharge is very free, we may use a
solution of Permanganate of Potash, one to three grains
to the ounce of water. The Acetous Tincture of Opium
is sometimes a good remedy in these cases, in the proportion of from f℈ij, to f℥ss, to ℥iv of water.

CHRONIC CONJUNCTIVITIS.—In this disease, we find that
the ordinary mild collyria may be used with considerable advantage. Thus I have obtained very good results from the use of Belladonna, and the acid Collyrium
heretofore named. The employment of the tonic infusions will also prove beneficial. Either of the tonic
or astringent formula, heretofore given, may be employed, as in Formulas No. 22, 23, and 24, lessening the
Aromatic Sulphuric Acid in the first two to f℥ss.

In granular conjunctivitis, these means are not so
available, though here they may be employed as adjuvants to the treatment, as the milder inhalations are
generally used.

RHEUMATIC OPHTHALMIA.—In this case Collyria have
been generally regarded as useless. But the atomized
fluid relieves irritation and soothes the inflamed organ.
I generally order—℞ Tincture of Aconite gtt. xx;
Tincture of Belladonna, f℈j; water, f℥iv: Mix. The
same means might be used in all internal inflamma-

tions, or even in iritis, though in the last, the use of a collyrium of Atropia will be found much better.

In other affections of the eyes where Collyria is used, the employment of atomized fluids may be substituted, special means need not be named in these cases, as they do not differ from those commonly employed.

THE EAR.

Of course the employment of the atomizer in diseases of the ear is but limited, being confined entirely to affections of the external auditory meatus. But even this limited action is sometimes available, as no diseases are more painful than those affecting this part.

Acute inflammation of the ear may be more speedily relieved by the use of narcotics in this way, than by any other means. Tinctures of Aconite, Stramonium, Belladonna, or Opium, added to water may be thus employed. I have used R.—Tincture Aconite, (root,) fʒij ; Tinct. Stramonium, fʒss; Water, fʒiijss. Mix.

In *otorrhœa*, we employ similar means to those we use as injections. The Tincture of Muriate of Iron, or solution of Perchloride of iron, fʒss to Water fʒiv; sometimes answers well. At other times I have used a solution of Permanganate of Potash, grs. ij to grs. v.; to Water fʒj. The astringent and tonic Formula may be employed in the same proportions they are used in diseases of the chest.

OTHER LOCAL APPLICATIONS.

Though it might be employed in any local affection, we will find that its use will be principally confined to the treatment of ulcers. For this purpose it sometimes

proves very valuable, though at others no better, if so
good, as other applications.   If used, the usual remedies
will be employed.

It has been used to a limited extent in neuralgic affections, but when any change from the common mode
was necessary, we would find the best means in the use
of the hypodermic syringe.

Some experiments have been made on *Diseases of the
Uterus*, but nothing definite can be said about its application here.   Whilst it would seem to be a desirable
method of applying remedies, the disagreebleness of its
employment, and limited action, will prevent its being
used to any considerable extent.   In painful affections
of the os uteri, as in ulceration and cancer, and in dys-
menorrhœa, it will doubtless be used to some extent.
The fact that a cylindrical speculum has to be used, and
that the fluid is therefore confined to a very small part,
is an insuperable objection to its use in other cases.

VI.

ON THE EMPLOYMENT OF ATOMIZED FLUIDS AS DISINFECTANTS, AND TO DESTROY ANIMAL AND VEGETABLE MIASMS.

An efficient means of disinfecting sick rooms, hospitals, prisons, ships, etc, has long been desired. We have disinfecting fluids in abundance, but they have never as yet been so applied as to be of much value.

I have determined by experiment that we may remove offensive odors, and destroy the germs of animal miasm, very speedily and with great certainty with the apparatus heretofore described. Thus taking the coal oil atomizer, Fig. 6, I have purified the air of a large and very foul dissecting room in an hour, and that with the use of but one ounce of the Solution of Chlorinated Soda. I have attained the same results by the employment of a solution of Sulphate of Iron, and the Sulphite of Soda. It is done so thoroughly that the room would not be offensive to any person. And as evidences that the odor was not simply covered up, we find that it is reproduced very slowly and only by renewed processes of decomposition.

With this apparatus, the air of a sick chamber can be thoroughly freed of all bad odors, and what is more to the purpose, the animal miasm upon which the spread of disease so frequently depends, is entirely destroyed. This is of very great importance, as the sick in some affections are continually poisoned by the decomposing emanations from their own bodies. Not only is it of advantage to the sick, but we thus prevent the spread of contagious diseases.

In seasons of epidemic disease, it will prove especially beneficial. The evidence has been growing stronger and stronger, until but few will deny that the spread of contagion depends to a very considerable extent upon an atmosphere already impure from animal and vegetable decomposition. It matters not whether it is typhoid fever, or cholera, or small-pox, measles, or scarlatina, or dysentery, the same rule holds good. During such epidemics, therefore, especial attention should be paid to the disinfecting of all places where decomposition is going on.

When the popular mind becomes alarmed by the approach of an epidemic, as of cholera, there is a spasmodic effort toward cleanliness, which is rather injurious than otherwise, in this—that a much larger and fresh surface is exposed to the action of the air; in fact the dirt is stirred up in place of being removed, and that at the time when it should remain at rest. I do not desire to advocate dirt, but I object decidedly to placing it in the most active condition possible during an epidemic. Do the cleaning before the epidemic comes, but when it is upon us, destroy the products of decomposition and keep it in as quiescent a condition as possible.

The apparatus represented in Fig. 6 will be sufficient

to thoroughly free a dwelling house from all unpleasant odors and animal miasms of which they are the sign. I prefer the use of the solution of Chlorinated Soda, but Chloride of Lime, or Sulphate of Iron may be employed with equal advantage. For more extended use in epidemics, I employ a much larger apparatus, using flexible hose, which can be carried wherever necessary.

It is very certain that the process of absorption is very rapid from the lungs, especially when the agent is in a gaseous form. Air becomes charged with moisture, and, in case of medicated fluids, with the medicine to a certain degree. May we not, therefore, suppose that in some cases agents might be introduced into the circulation with advantage? I have experimented in this line to some extent, but hardly like to express an opinion as yet.

• Should this be the fact, it is barely possible that we may thus find an antidote to the blood poisons, as in cholera, typhoid and typhus fevers, the eruptive fevers, etc. I think the subject is worthy of investigation and would be glad to learn of any investigations in this direction.

8

INDEX.

A

B

C

D

www.ingramcontent.com/pod-product-compliance
Lightning Source LLC
Chambersburg PA
CBHW022104210326
41519CB00056B/1201